Topics in Applied Physics Volume 42

Topics in Applied Physics Founded by Helmut K. V. Lotsch

1 **Dye Lasers** 2nd Ed. Editor: F. P. Schäfer
2 **Laser Spectroscopy** of Atoms and Molecules. Editor: H. Walther
3 **Numerical and Asymptotic Techniques in Electromagnetics** Editor: R. Mittra
4 **Interactions on Metal Surfaces** Editor: R. Gomer
5 **Mössbauer Spectroscopy** Editor: U. Gonser
6 **Picture Processing and Digital Filtering** 2nd Edition. Editor: T. S. Huang
7 **Integrated Optics** 2nd Ed. Editor: T. Tamir
8 **Light Scattering in Solids** Editor: M. Cardona
9 **Laser Speckle** and Related Phenomena Editor: J. C. Dainty
10 **Transient Electromagnetic Fields** Editor: L. B. Felsen
11 **Digital Picture Analysis** Editor: A. Rosenfeld
12 **Turbulence** 2nd Ed. Editor: P. Bradshaw
13 **High-Resolution Laser Spectroscopy** Editor: K. Shimoda
14 **Laser Monitoring of the Atmosphere** Editor: E. D. Hinkley
15 **Radiationless Processes** in Molecules and Condensed Phases. Editor: F. K. Fong
16 **Nonlinear Infrared Generation** Editor: Y.-R. Shen
17 **Electroluminescence** Editor: J. I. Pankove
18 **Ultrashort Light Pulses** Picosecond Techniques and Applications Editor: S. L. Shapiro
19 **Optical and Infrared Detectors** 2nd Ed. Editor: R. J. Keyes
20 **Holographic Recording Materials** Editor: H. M. Smith
21 **Solid Electrolytes** Editor: S. Geller
22 **X-Ray Optics.** Applications to Solids Editor: H.-J. Queisser
23 **Optical Data Processing.** Applications Editor: D. Casasent
24 **Acoustic Surface Waves** Editor: A. A. Oliner
25 **Laser Beam Propagation in the Atmosphere** Editor: J. W. Strohbehn
26 **Photoemission in Solids I.** General Principles Editors: M. Cardona and L. Ley
27 **Photoemission in Solids II.** Case Studies Editors: L. Ley and M. Cardona
28 **Hydrogen in Metals I.** Basic Properties Editors: G. Alefeld and J. Völkl
29 **Hydrogen in Metals II** Application-Oriented Properties Editors: G. Alefeld and J. Völkl
30 **Excimer Lasers** Editor: Ch. K. Rhodes
31 **Solar Energy Conversion.** Solid-State Physics Aspects. Editor: B. O. Seraphin
32 **Image Reconstruction from Projections** Implementation and Applications Editor: G. T. Herman
33 **Electrets** Editor: G. M. Sessler
34 **Nonlinear Methods of Spectral Analysis** Editor: S. Haykin
35 **Uranium Enrichment** Editor: S. Villani
36 **Amorphous Semiconductors** Editor: M. H. Brodsky
37 **Thermally Stimulated Relaxation in Solids** Editor: P. Bräunlich
38 **Charge-Coupled Devices** Editor: D. F. Barbe
39 **Semiconductor Devices** for Optical Communication Editor: H. Kressel
40 **Display Devices** Editor: J. I. Pankove
41 **The Computer in Optical Research** Methods and Applications Editor: B. R. Frieden
42 **Two-Dimensional Digital Signal Processing I.** Linear Filters Editor: T. S. Huang
43 **Two-Dimensional Digital Signal Processing II.** Transforms and Median Filters. Editor: T. S. Huang
44 **Turbulent Reacting Flows** Editors: P. A. Libby and F. A. Williams
45 **Hydrodynamic Instabilities and the Transition to Turbulence** Editors: H. L. Swinney and J. P. Gollub
46 **Glassy Metals I** Editors: H.-J. Güntherodt and H. Beck
47 **Sputtering by Particle Bombardment I** Editor: R. Behrisch
48 **Optical Information Processing** Fundamentals Editor: S. H. Lee

Two-Dimensional Digital Signal Processing I

Linear Filters

Edited by T. S. Huang

With Contributions by
L.T. Bruton   T. S. Huang   R. M. Mersereau
B. T. O'Conner   P. A. Ramamoorthy   J. W. Woods

With 77 Figures

Springer-Verlag Berlin Heidelberg GmbH 1981

Professor *Thomas S. Huang*, Ph.D.
Department of Electrical Engineering and
Coordinated Science Laboratory
University of Illinois
Urbana, IL 61801, USA

ISBN 978-3-662-30894-3 ISBN 978-3-540-38443-4 (eBook)
DOI 10.1007/978-3-540-38443-4

Library of Congress Cataloging in Publication Data. Main entry under title: Two-dimensional digital signal processing. (Topics in applied physics; v. 42) Includes bibliographies and index. 1. Signal processing – Digital techniques. I. Huang, Thomas S., 1936– II. Bruton, Len T. TK5102.5.T93 621.38′043 80-25511

Originally published by Springer-Verlag Berlin Heidelberg New York in 1981
Softcover reprint of the hardcover 1st edition 1981

2153/3130-543210

Preface

Activities in digital image processing have been increasing rapidly in the past decade. This is not surprising when one realizes that in a broad sense image processing means the processing of multidimensional signals and that most signals in the real world are multidimenional. In fact, the one-dimensional signals we work with are often collapsed versions of multidimensional signals. For example, speech is often considered as a one-dimensional signal, viz., a function of a single variable (time). However, speech originally exists in space and therefore is a function of 4 variables (3 spatial variables and time).

There are analog (optical, electrooptical) as well as digital techniques for image processing. Because of the inherent advantages in digital techniques (flexibility, accuracy), and because of the rapid progress in computer and related technologies, such as LSl, and VLSI, it is fair to say that except for some specialized problems, digital techniques are usually preferred.

The purpose of this book and its companion on transforms and median filters is to provide in-depth treatments of three of the most important classes of digital techniques for solving image processing problems: linear filters, transforms, and median filtering. These two books are related but can be used independently.

In an earlier volume of this series, *Picture Processing and Digital Filtering* (first edition, 1975), selected topics in two-dimensional digital signal processing including transforms, filter design, and image restoration, were treated in depth. Since then, a tremendous amount of progress has been made in these areas. In 1978 when we were planning on a second edition of that book (published in 1979), a decision was made not to make significant revisions, but only to add a brief new chapter surveying the more recent results. We projected that in-depth treatments of some of the important new results would appear in future volumes of the Springer physics program.

The present treatment on two-dimensional digital signal processing is divided into three parts. In the first part on linear filters, which is contained in this book, major recent results in the design of two-dimensional nonrecursive and recursive filters, stability testing, and Kalman filtering (with applications to image enhancement and restoration) are presented. Among the highlights are the discussions on the design and stability testing of half-plane recursive filters, a topic of great current interest.

The second and the third parts are contained in the subsequent book. In the second part on transforms, two topics are discussed: algorithms for transposing large matrices, and number-theoretic techniques in transforms and convolution. Here a detailed derivation of the Winograd Fourier transform algorithm is given.

In the first and the second parts, the main concern is linear processing. In the third part on median filtering, a particular nonlinear processing technique is studied. Median filtering has become rather popular in image and speech processing. However, published results on it have been scarce. The two chapters of the third part contain new results most of which will be published here for the first time.

The chapters are tutorial in nature, yet they bring the readers to the very forefront of current research. The treatment will be useful as a reference book for working scientists and engineers, and as a supplementary textbook for regular or short courses on digital signal processing, image processing, and digital filtering.

Urbana, Illinois *Thomas S. Huang*
September 1980

Contents

Contributors

Bruton, Leonard T.
Department of Electrical Engineering
University of Calgary, 2500 University Drive N.W.
Calgary, Alberta, Canada T2N 1N4

Huang, Thomas S.
Department of Electrical Engineering and Coordinated
Science Laboratory, University of Illinois
Urbana, IL 61801, USA

Mersereau, Russell M.
School of Electrical Engineering, Georgia Institute of Technology
Atlanta, GA 30332, USA

O'Conner, Brian T.
TRW Defense Space System Group (DSSG), One Space Park
Redondo Beach, CA 90278, USA

Ramamoorthy, Panapakkam A.
School of Engineering, Western New England College
Springfield, MA 01119, USA

Woods, John W.
Electrical and Systems Engineering Department
Rensselaer Polytechnic Institute
Troy, NY 12181, USA

1. Introduction

T. S. Huang

1.1 Image Processing

The field of application which motivates the mathematical techniques presented in this book is image processing. To provide the proper perspective, we start by giving an overview of the three major areas of image processing: efficient coding, restoration and enhancement, and pattern recognition.

1.1.1 Efficient Coding

In the digital transmission and storage of images, the amounts of bits required are often astronomical. Efficient coding to reduce the requirements is not only desirable but also in many cases mandatory.

Much past research in coding moving images was motivated by videophones. More recently, the applications of interest are broadcast TV, and teleconferencing. If we sample the NTSC color video signal (broadcast TV) at 10.7 MHz (3 times the color subcarrier frequency), and quantize each sample to 256 levels (8 bits), we obtain a bit rate of 85.6 Mbits/s. By using various intra- and inter-frame coding techniques, broadcast-quality systems have been constructed with bit rates of around 20 Mbits/s. For teleconferencing, less motion is involved and more degradation can be tolerated, so that bit rates of around 2–5 Mbits/s have been achieved.

Another problem where efficient coding is needed is the storage and transmission of multispectral LANDSAT images. It is estimated that each of LANDSATS 1, 2, and 3 creates approximately 1.5×10^{13} bits/yr. LANDSAT D with its higher resolution images will create approximately 3.7×10^{15} bits/yr. A large compression factor is needed to facilitate the data handling. Since different users of these multispectral images are to extract different types of information from them, it is difficult to decide what degradations of the data are acceptable. To be on the safe side, one will probably demand essentially no degradation from the original digitized data. This makes large compression impossible.

For both intra- and inter-frame coding, the effective and currently popular techniques are dithering, predictive coding, and transform coding. In many systems, these techniques are used in combination.

1.1.2 Restoration and Enhancement

Both restoration and enhancement aim at improving image quality. The demarcation between the two is not clear. Roughly speaking, when we talk about restoration, we have some definite degradations (e.g., defocusing) in mind and we want to remove these degradations to obtain an ideal image (an image which we would have obtained in the absence of these degradations). Enhancement, on the other hand, is broader in scope. Here, we want to put the image in a form that is suitable for our purpose. We do not necessarily want the ideal image. For example, we may want to oversharpen the edges, or to use false colors.

Perhaps the most prominent example of successful image restoration is the work done at the Jet Propulsion Laboratory with images of the Moon, Mars, and other planets taken by TV cameras on board of artificial satellites. They have succeeded in compensating for various image degradations including: random noise, interference, geometrical distortion, field nonuniformity, contrast loss, and blurring. Another problem which has spurred much of the research in image restoration, but to which satisfactory solutions are yet to be found, is imaging through the atmosphere. The problem is to obtain good images of planets, stars, and artificial satellites by imaging systems based on earth. The dominating degradation in this case is the blurring due to atmospheric turbulence.

There are two approaches to image restoration: "a priori" and "a posteriori". In the a priori approach, one tries to invent novel imaging schemes which would give good images to start with. In the a posteriori approach, one uses conventional imaging systems such as a camera, and then attempts to improve the image thus obtained. In the case of imaging through the atmosphere, several a priori methods have been proposed. These include holographic techniques and coded-aperture imaging – the latter has achieved some success. An area of intensive current research is adaptive optics. Here, the phase disturbance due to the atmosphere is measured, and controllable deformable optics are used to compensate for the disturbance in real time.

In the a posteriori approach, most of the effort has been concentrated on linear degradations. This is because of the fact that most image degradations can be modeled adequately by a cascade of three subsystems: a linear part, a nonlinear no-memory part, and noise. For example, such a model can be used for film, where the linear part represents blurring due to chemical diffusion, the nonlinear no-memory part represents the H–D curve, and the noise is due to film grains. The term "no-memory" is used in the following sense. A two-dimensional system is said to have no memory, if the value of the output at any given point depends only on the value of the corresponding point in the input but not on the values of any other input points. If the characteristics of the degradations are known, then compensation for the nonlinear no-memory part is straightforward. The more difficult task is to compensate for or equivalently to find the inverse of the linear degradation. The performance of image restoration methods is ultimately limited by the presence of noise.

If the linear degradation is spatially invariant, then Fourier techniques (e.g., Wiener inverse filtering) can be used to do the inversion. Examples of linear spatially invariant degradations include defocusing and spherical aberration of the lens, and translational motion of the camera relative to the object. In the Fourier techniques, one applies a linear spatially invariant (LSI) inverse filter to the degraded image to obtain the restoration. LSI inverse filters have severe limitations. Since such a filter multiplies each spatial frequency component of the degraded image by a complex constant, it cannot alter the signal-to-noise ratio at any individual spatial frequency. The only thing it can do is to boost up those frequency components where the signal is larger than the noise and suppress those frequency components where the noise is larger than the signal. A more dramatic implication is that there is no way for such a filter to recover those spatial-frequency components of the ideal image which were cut off by a band-limiting degrading system – this fact is sometimes stated as: "LSI inverse filters cannot achieve superresolution".

In order to achieve superresolution, one has to use nonlinear (NL) or linear spatially varying (LSV) techniques, and have additional information about the signal. For noncoherent imaging systems one piece of information we always have: the image is nonnegative. Another piece of information which is not always true is that the scene giving rise to the image may consist of a spatially limited object lying in a uniform background.

A number of LSV and NL restoration methods have been developed which make use of these pieces of information. They fall into two categories: frequency-domain and spatial-domain techniques. The latter have been proven more successful mainly because the information about the signal is stated in the spatial domain and it is very awkward to try to translate it into the frequency domain. The spatial-domain techniques have the further advantage that they can be applied to linear spatially varying degradations (such as lens aberrations, except spherical aberration, and rotational motion of the camera relative to the object) as well.

Among LSV techniques, several are based on generalized matrix inverses. These include singular-value decomposition, and the projection iterative method. NL techniques include maximum-entropy methods and Bayes estimation. Generally, these methods can achieve much better restoration than Fourier methods. However, one has to pay the price of computational complexity. For example, applying any of these spatial-domain methods to even a small image (128 × 128 points, say) can take several hours on a minicomputer (e.g., PDP 11/45 with floating-point hardware).

1.1.3 Pattern Recognition

In a narrow sense, pattern recognition means the classification of a given unknown pattern into one of a number of standard classes. This is normally done in two steps. First, features are extracted from the given pattern. These features are represented by numbers. The vector with these features as components is called the feature vector. Second, based on the feature vector a classification is made. Usually, training samples are used to design the classifier.

The oldest area of application of pattern recognition is probably optical character recognition (OCR). More recently, pattern recognition has found successful applications in remote sensing (e.g., land use classification of LANDSAT images), biomedical problems (e.g., classification of white blood cells), and industrial problems (e.g., quality control).

The most interesting recent development is to view pattern recognition in a broad sense as scene analysis. The problem is to develop automatic and interactive systems which will collect sensor and nonsensor data and information and come up with a description of the scene. To solve the problem, one needs tools from diverse disciplines, chief among them are multidimensional signal processing, pattern recognition (in the narrow sense), and artificial intelligence.

1.1.4 Outline

As the reader has no doubt gathered from the above discussions, image processing is an extremely broad field. The techniques required in solving image processing problems are many and various. In this volume, we shall concentrate on linear filters. Transforms and median filters are treated in the subsequent volume [1.1].

Linear filters are useful in image restoration and enhancement, in prewhitening for spectral analysis, and in matched filtering for pattern detection. Transforms and related methods enable computationally efficient implementations of linear filters. More detailed discussions on the topics contained in the present book are given in the following sections.

1.2 Linear Filters

In [Ref. 1.2, Chaps. 3 and 4], results in the design of two-dimensional nonrecursive filters, and the design and stability testing of two-dimensional recursive filters prior to 1975 were discussed. Since then much progress has been made. Some of the important new results are presented in detail in Chaps. 2–4. In addition, the topic of two-dimensional Kalman filtering which was mentioned only in passing near the end of [Ref. 1.2, Chap. 1] is here given a whole chapter (Chap. 5). For general background material, the reader is referred to [1.3–5].

1.2.1 Nonrecursive Filter Design

In [Ref. 1.2, Chap. 3], four methods of designing nonrecursive two-dimensional filters were described: window technique, frequency sampling, conventional linear programming, and iterative linear programming. Since around 1975, two new good design techniques have emerged. The first is a transformation method proposed by *McClellan* [1.6] and later generalized by *Mersereau* and *Mecklenbrauker* [1.7, 8]. In this method, one designs two-dimensional nonrecursive filters by applying a change of variables to optimum minimax one-dimensional filters. It is especially suitable for designing circularly symmetric low-pass and band-pass filters, and more generally filters with quadrilateral symmetry. The method is computationally easy and in many cases gives optimal designs.

The second method is the extension of the Remez multiple exchange algorithm to the design of two-dimensional nonrecursive filters [1.9]. This is probably the best general purpose design tool for nonrecursive filters with Chebyshev error norm (minimax).

In Chap. 2 by Mersereau, the transformation method and the optimization approach to two-dimensional nonrecursive filter design are treated in depth. The window method is also reviewed briefly. We call the reader's attention to a number of papers on windows [1.10–12] as well as a paper by *Manry* and *Aggarwal* [1.13] which describes a method of reducing the order of impulse response by using nonrectangular arrays (i.e., arrays where the nonzero elements do not form rectangles).

1.2.2 Recursive Filters–Stability

A key problem in designing two-dimensional recursive filters is the test of stability. Many earlier results on stability were discussed in detail in [Ref. 1.2, Chap. 4]. For more recent developments on this and related problems, the reader is referred to the masterful review paper by *Jury* [1.14] and [1.15–21]. Recently, based on the pioneering work of *Pistor* [1.22], *Dudgeon* [1.23], and *Ekstrom* and *Twogood* [1.24], *O'Connor* and *Huang* developed some very efficient methods of testing the stability of two-dimensional recursive filters [1.25–27].

In [Ref. 1.2, Chap. 4], it was conjectured that the double least squares inverse method would stabilize an unstable filter. It has been demonstrated since then that this conjecture is not true in general [1.28]. However, it is interesting to investigate under what conditions the conjecture is valid [1.29–37].

In many stability tests, bilinear transformations are used. As *Goodman* [1.38] and *Bickart* [1.39] pointed out, this may cause trouble because of singularities of the second kind. Recently, *Reddy* et al. developed a method of generating two-dimensional digital functions without nonessential singularities of the second kind [1.40].

Almost all existing stability tests are done in the frequency domain. However, *Garibotto* [1.41] has presented a spatial-domain technique based on the analysis of the impulse response of an all-pass phase filter associated with the denominator polynomial of the given filter transfer function. *Kamen* [1.42] has developed a theory of asymptotic stability for a large class of half-plane two-dimensional digital filters. It is based on a spatial-domain representation consisting of a one-dimensional difference equation with coefficients in an algebra of one-dimensional functions.

Many important results on two-dimensional stability are discussed briefly in Chap. 3 and in detail in Chap. 4. In Chap. 4, the stability of a general class of two-dimensional recursive filters is treated. This class includes quarter-plane and nonsymmetrical half-plane filters as special cases.

1.2.3 Recursive Filters – Design Techniques

In [Ref. 1.2, Chap. 4], three methods of designing two-dimensional recursive filters were discussed: separable filters, rotation and shifting of one-dimensional filters, and spatial-domain least-square approximation.

The approach of designing two-dimensional recursive filters by rotating one-dimensional filters has been studied extensively by *Costa* and *Venetsanopoulis* [1.43, 44]. An extension was discussed in [1.45]. The idea of separable-sum approximation [1.46–48] was pursued further by *Takahashi* and *Tsuji* using Laguerre functions [1.49]. Related papers are [1.50, 51]. Several design methods based on optimization techniques have been developed [1.52–55]. In particular, the method of *Aly* and *Fahmy* [1.55] uses l_p approximations of magnitude and group delay, and guarantees stability through the use of a frequency transformation [1.56]. *Ramamoorthy* and *Bruton* [1.56, 57] proposed a method based on the decomposition of an $n \times n$ multivariable positive real admittance matrix and used Koga's result that all such matrices can be realized as multivariable finite lumped passive networks. The method guarantes stability. Several design techniques impose relatively simple sufficient conditions for stability, which however are not necessary. These include the methods described in [1.58, 59]. Other design techniques can be found in [1.60, 61].

All the design techniques mentioned above are for quarter-plane filters, i.e., the support of the impulse response of the filter is a quarter-plane. The design is more flexible, if we can extend the support of the impulse response to a half-plane. The design of half-plane filters is in its infancy [1.62–67] and an important area for further research.

For analyses of quantization errors in two-dimensional digital filters, see [1.8, 68–77]. For two-dimensional filter structures, see [1.78–80]. *Mersereau* [1.81] discussed filtering and Fourier transformation of hexagonally sampled signals. In [Ref. 1.2, Chap. 4], a state variable implementation of two-dimensional recursive filters was briefly discussed. For related papers, see [1.82–84]. This topic is of course closely related to two-dimensional Kalman filters, the subject matter of Chap. 5.

In Chap. 3 the design of both quarter-plane and nonsymmetrical half-plane filters is discussed. The highlight of the chapter is undoubtedly the method of

Ramamoorthy and *Bruton* for designing quarter-plane filters as mentioned earlier and its extension to half-plane filter design.

1.2.4 Kalman Filtering

An obvious way of noise reduction is to use Wiener (linear least-squares) filters. An alternative is Kalman filters. *Habibi* and co-workers were among the first who suggested the extension of Kalman filtering to two dimensions for image noise reduction [1.85–89]. More recently, important contributions have been made by *Jain* and *Woods* [1.89–92]. In particular, *Woods* has proposed several ways of making the computation more efficient, see also [1.93–97].

In Chap. 5 various aspects of two-dimensional Kalman filtering are discussed in depth. The emphasis is on modeling, computational algorithms, and system identification.

References

1.1 T.S.Huang (ed.): *Two-Dimensional Digital Signal Processing II: Transforms and Median Filters*, Topics in Applied Physics, Vol. 43 (Springer, Berlin, Heidelberg, New York 1980)

1.2 T.S.Huang (ed.): *Picture Processing and Digital Filtering*, 2nd ed, Topics in Applied Physics, Vol. 6 (Springer, Berlin, Heidelberg, New York 1979)

1.3 S.K.Mitra, M.P.Ekstron (eds.): *Two-Dimensional Digital Signal Processing* (Dowden, Hutchinson, and Ross, Stroudsburg, Pennsylvania 1978)

1.4 N.K.Bose (ed.): Special Issue on Multidimensional Systems, Proc. IEEE **65** (June, 1977)

1.5 R.M.Mersereau, D.E.Dudgeon: Proc. IEEE **63**, 610–623 (1975)

1.6 J.H.McClellan: "The Design of Two-Dimensional Digital Filters by Transformation", Proc. 7th Ann. Princeton Conf. Information Science and Systems (1973) pp. 247–251

1.7 R.M.Mersereau, W.F.G.Mecklenbrauker, T.F.Quatieri, Jr.: IEEE Trans. CS-**23**, 405–411 (1976)

1.8 W.F.G.Mecklenbrauker, R.M.Mersereau: IEEE Trans. CS-**23**, 414–422 (1976)

1.9 D.B.Harris, R.M.Mersereau: "A Comparison of Iterative Methods for Optimal 2-d Filter Design", Proc. Intern. Conf. Acoustics, Speech, and Signal Processing (ASSP) (May 1977) p. 527

1.10 F.J.Harris: Proc. IEEE **66**, 51–83 (1978)

1.11 N.C.Geckinli, D.Yavuz: IEEE Trans. ASSP-**26**, 501–507 (1978)

1.12 A.Blomquist: Proc. IEEE **67**, 438–439 (1979)

1.13 M.T.Manry, J.K.Aggarwal: IEEE Trans. ASSP-**26**, 314–318 (1978)

1.14 E.I.Jury: Proc. IEEE **66**, 1018–1047 (1978)

1.15 N.K.Bose: Proc. IEEE **65**, 824–840 (1977)

1.16 N.K.Bose, S.Basu: IEEE Trans. CS-**25**, 684–693 (1978)

1.17 N.K.Bose: IEEE Trans. ASSP-**27**, 1–4 (1979)

1.18 E.Zeheb, E.Walach: "Zero Sets of Multiparameter Functions and Stability of Multidimensional Systems", IEEE Trans. ASSP (to be published)

1.19 M.G.Strintzis: IEEE Trans. CS-**24**, 432–437 (1977)

1.20 R.Decarlo, R.Saeks, J.Murray: Proc. IEEE **65**, 978–979 (1977)

1.21 G.A.Shaw: "An Algorithm for Testing Stability of Two-Dimensional Digital Recursive Filters", Proc. IEEE Intern. Conf. ASSP (April 1978) pp. 769–772

1.22 P.Pistor: IBM J. Res. Dev. **18**, 59–71 (1974)

1.23 D.Dudgeon: IEEE Trans. ASSP **25**, 476–484 (1977)
1.24 M.Ekstrom, R.Twogood: "A Stability Test for Two-Dimensional Recursive Digital Filters Using the Complex Cepstrum", Proc. 1977 Intern. Conf. ASSP (April 1977) pp. 535–538
1.25 B.T.O'Connor, T.S.Huang: "Stability of General Two-Dimensional Recursive Digital Filters", Tech. Rpt. TR-EE 77-36, School of Electrical Engineering, Purdue University (Oct. 1977)
1.26 B.T.O'Connor, T.S.Huang: "Techniques for Determining the Stability of Two-Dimensional Recursive Filters and Their Applications to Image Restoration, Tech. Rpt. TR-EE 78-18 (May 1978), School of Electrical Engineering Purdue University
1.27 B.T.O'Connor, T.S.Huang: IEEE Trans. ASSP-**26**, 550–560 (1978)
1.28 Y.Genin, Y.Kamp: Electron. Lett. **11**, 330–331 (1975)
1.29 Y.Genin, Y.Kamp: Proc. IEEE **65**, 873–881 (1977)
1.30 Y.V.Genin, Y.G.Kamp: IEEE Trans. ASSP-**25**, 92–93 (1977)
1.31 P.Delsarte, Y.Genin, Y.Kamp: Proc. IEEE **66**, 1162 (1978)
1.32 P.Delsarte, Y.V.Genin, Y.G.Kamp: IEEE Trans. CS-**26**, 59–66, 898 (1979)
1.33 J.B.Bednar: IEEE Trans. ASSP-**23**, 583–585 (1975)
1.34 E.I.Jury: "A Review of Shank's Conjecture and Comments on Its Validity", in Proc. Asilomar Conf. on Cir. Syst. and Computers (Nov. 1976)
1.35 B.D.O.Anderson, E.I.Jury: IEEE Trans. ASSP-**24**, 574–575 (1976)
1.36 E.I.Jury, V.R.Kolavennu, B.D.O.Anderson: Proc. IEEE **65**, 887–892 (1977)
1.37 E.I.Jury, L.F.Chaparro: Proc. IEEE **67**, 1444–1445 (1979)
1.38 D.Goodman: Proc. IEEE **66**, 796–797 (1978)
1.39 T.A.Bickart: Proc. IEEE **66**, 983 (1978)
1.40 C.H.Reddy, P.K.Rajan, M.N.S.Swamy, V.Ramachandran: "Generation of Two-Dimensional Digital Functions Without Nonessential Singularities of the Second Kind", IEEE Intern. Conf. ASSP, Washington, D.C. (April 2–4, 1979)
1.41 G.Garibotto: Spatial-Domain Stability Test, presented at the 1st ASSP workshop on 2D-Digital Signal Processing, Berkeley, CA., (Oct. 3–4, 1979)
1.42 E.W.Kamen: "Asymptotic Stability of linear Shift-Invariant, Two-Dimensional Digital Filters", preprint (Dec. 1979)
1.43 J.M.Costa, A.N.Venetsanopoulis: IEEE Trans. ASSP-**22**, 432–443 (1974)
1.44 J.M.Costa, A.N.Venetsanopoulis: Can. Electron. Eng. **4**, 33–40 (1979)
1.45 D.M.Goodman: IEEE Trans. ASSP-**26**, 290–304 (1978)
1.46 S.Treitel, J.L.Shanks: IEEE Trans. GE-**9**, 10–27 (1971)
1.47 T.S.Huang, W.F.Schreiber, O.J.Tretiak: Proc. IEEE **59**, 1588–1609 (1971)
1.48 S.Chakrabarti, N.K.Bose, S.K.Mitra: J. Franklin Inst. **229**, 53–66 (1975)
1.49 S.Takahashi, S.Tsuji: Trans. Inst. Electron. Communication Eng. (Jp) **60**, No. 6, 9–16 (1977)
1.50 J.F.Abramatic, F.Germain, E.Rosencher: IEEE Trans. ASSP-**27**, 445–453 (1979)
1.51 K.Hirano, J.K.Aggarwal: IEEE Trans. CS-**25**, 1066–1076 (1978)
1.52 D.E.Dudgeon: IEEE Trans. ASSP-**23**, 264–267 (1975)
1.53 G.A.Maria, M.M.Fahmy: IEEE Trans. ASSP-**22**, 15–21 (1974)
1.54 G.A.Maria, M.M.Fahmy: IEEE Trans. CS-**21**, 431–436 (1974)
1.55 A.H.Aly, M.M.Fahmy: IEEE Trans. CS-**25**, 908–916 (1978)
1.56 P.A.Ramamoorthy, L.T.Bruton: "Frequency Domain Approximation of Stable Multidimensional Discrete Recursive Filters', Proc. Intern. Symp. Circuits and Systems (April 1977)
1.57 P.A.Ramamoorthy, L.T.Bruton: "Design of Stable Two-Dimensional Digital Recursive Filters and Applications in Image Processing", Proc. 20th Midwest Symp. on Circuits and Systems (August 1977)
1.58 A.Chottera, G.A.Jullien: "Designing Near Linear Phase Recursive Filters Using Linear Programming", Proc. IEEE Intern. Conf. on ASSP (May 1977) pp. 88–92
1.59 A.M.Ali: "Design of Inherent Stable Two-Dimensional Recursive Filters Imitating the Behavior of One-Dimensional Analog Filters", Proc. IEEE Intern. Conf. ASSP (April 1978) pp. 765–768

1.60 M.Lal: "Space-Domain Design of Two-Dimensional Recursive Digital Filters", Proc. IEEE Symp. Circuits and Systems, Newton, Mass. (March 1975)
1.61 N.A.Pendergrass, S.K.Mitra, E.I.Jury: IEEE Trans. CS-**23**, 26–35 (1976)
1.62 M.P.Ekstrom, R.E.Twogood, J.W.Woods: "Design of Stable Two-Dimensional Half-Plane Recursive Filters Using Spectral Factorization", IEEE Intern. Conf. ASSP, Tulsa, Oklahoma (April 1978)
1.63 M.P.Ekstrom, R.E.Twogood: "Recursive Filter Design by Spectral Factorization", presented at the 1st ASSP workshop on 2D Digital Signal Processing, Berkely, CA (Oct. 3–4, 1979)
1.64 G.A.Shaw, R.M.Mersereau: Proc. IEEE **67**, 1081–1083 (1979)
1.65 P.K.Rajan, M.N.S.Swamy: IEEE Trans. ASSP-**27**, 506–511 (1979)
1.66 H.Chang, J.K.Aggarwal: IEEE Trans. CS-**25**, 1051–1059 (1978)
1.67 H.Chang, J.K.Aggarwal: "Fundamentals of Two-Dimensional Recursive Filtering", in *Digital Signal Processing*, ed. by J. K. Aggarwal (Western Periodicals Company, North Hollywood, CA 1979)
1.68 G.A.Maria, M.M.Fahmy: IEEE Trans. CS-**22**, 826–830 (1975)
1.69 K.O.Shipp, Jr., J.K.Aggarwal: IEEE Trans. ASSP-**24**, 339–341 (1976)
1.70 M.-D.Ni, J.K.Aggarwal: IEEE Trans. C-**25**, 755–759 (1976)
1.71 T.-L.Chang: IEEE Trans. CS-**24**, 15–19 (1977)
1.72 P.Agathoklis, E.I.Jury, M.Mausour: "A Note on the Evaluation of Quantization Error in Two-Dimensional Digital Filters", preprint (Oct. 1978)
1.73 A.Fettweis: IEEE Trans. CS-**25**, 1060–1066 (1978)
1.74 N.K.Bose, J.P.Guiver: "Multivariate Polynomial Positivity Invariance Under Coefficient Perturbation with Applications", presented at First workshop on 2-D Digital Signal Processing, Berkeley, CA (October 3–5, 1979)
1.75 G.Shaw: "Effects of Coefficient Quantization on Stability", presented at the 1st ASSP workshop on 2-D Digital Signal Processing, Berkeley, CA., (Oct. 3–4, 1979)
1.76 N.G.El-Agizi, M.M.Fahmy: IEEE Trans. ASSP-**27**, 465–469 (1979)
1.77 S.H.Mneney, A.N.Venetsanopoulos, J.M.Costa: "The Effects of Quantization Errors on Rotated Filters", IEEE Trans. CS (to be published)
1.78 D.Harris: "Design of Symmetrical Half-Plane Filters", presented at the 1st ASSP workshop on 2D Digital Signal Processing, Berkeley, CA. (Oct. 3–4, 1979)
1.79 S.K.Mitra, S.Chakrabarti: IEEE Trans. ASSP-**26**, 544–550 (1978)
1.80 A.Fettweis: "Principles of Multidimensional Wave Digital Filtering", in *Digital Signal Processing*, ed. by J. K. Aggarwal (Western Periodicals Company, North Hollywood, CA 1979)
1.81 R.M.Mersereau: Proc. IEEE **67**, 930–940 (1979)
1.82 S. Y.Kung, B.C.Levy, M.Morf, T.Kailath: Proc. IEEE **65**, 945–961 (1977)
1.83 R.Eising: Proc. IEEE **67**, 866–868 (1979)
1.84 E.D.Sontag: IEEE Trans. ASSP-**26**, 480–482 (1978)
1.85 A.Habibi: Proc. IEEE **60**, 878–883 (1972)
1.86 M.G.Strintzis: Proc. IEEE **64**, 1255–1257 (1976)
1.87 N.E.Nahi, T.Assefi: IEEE Trans. C-**21**, 734–738 (1972)
1.88 N.E.Nahi, C.A.Franco: IEEE Trans. C-**21**, 305–311 (1973)
1.89 A.K.Jain, E.Angel: IEEE Trans. C-**23**, 470–476 (1976)
1.90 J.W.Woods, C.H.Radewan: "Reduced Update Kalman Filter – A Two-Dimensional Recursive Processor", Proc. Johns Hopkins Conf. Inf. Sci. Syst. (March 1976)
1.91 J.W.Woods, C.H.Radewan: IEEE Trans. IT-**23**, 473–482 (1977)
1.92 J.W.Woods, V.K.Ingle, A.Radpour, H.Kaufman: "Recursive Estimation with Non-Homogeneous Image Models", Proc. IEEE Conf. Pattern Recognition and Image Processing, Chicago (May 1978)
1.93 E.Wong: IEEE Trans. IT-**24**, 50–59 (1978)
1.94 B.R.Suresh, B.A.Shenoi: Comput. Graphics Image Processing **11**, 101–110 (1979)
1.95 T.Katayama, Y.Fujii: Int. J. Systems Sci. **9**, 543–562 (1978)
1.96 T.Katayama, M.Kosaka: IEEE Trans. AC-**24**, 130–132 (1979)
1.97 T.Katayama: "Restoration of Noisy Images Using a Two-Dimensional Linear Model", IEEE Trans. Syst., Man, and Cyber. (to be published)

2. Two-Dimensional Nonrecursive Filter Design

R. M. Mersereau

With 11 Figures

In this chapter the design of two-dimensional (2-D) nonrecursive digital filters will be considered. These filters share many characteristics with their one-dimensional counterparts. Like 1-D nonrecursive filters, for example, 2-D nonrecursive filters are always stable and they can have a frequency response function which is purely real. In addition, they can be implemented directly using a convolution sum or they can be implemented indirectly using discrete transforms. The design algorithms for 2-D nonrecursive filters will also be seen to be closely related to 1-D algorithms. On the other hand, there are differences which arise with the increase in dimensionality. These are due primarily to two factors. First, multidimensional designs possess more degrees of freedom than one-dimensional ones and this can be exploited. Secondly, the available mathematics for higher dimensional problems is more restrictive.

In this chapter attention will be focused on three methods for nonrecursive filter design – windows, transformations, and Chebyshev approximations. These are certainly not the only methods available, nor are they necessarily the best algorithms for all applications, but they are among the most popular design approaches. These three procedures also nicely illustrate the flavor of most multidimensional design methods and highlight some of the difficulties as well. The window and transformation procedures are straightforward; equiripple approximations are somewhat more involved.

2.1 2-D Nonrecursive Digital Filters

2.1.1 Filter Specifications

A (single input, single output) 2-D *digital filter* is a linear, shift-invariant transformation which acts on one doubly indexed array, $x(m,n)$, known as the input, to produce another doubly indexed array, $y(m,n)$, known as the output. A 2-D *nonrecursive* filter is one for which the sample of the output at (m,n) can be computed as a linear combination of a finite number of samples of x. Thus for a nonrecursive filter

$$y(m,n)=\sum_{k}\sum_{l}h(k,l)x(m-k,n-l)\ , \tag{2.1}$$

where the summation encompasses only a finite number of terms. The set of points $\{(k, l)\}$ which are included in this sum are referred to as the *region of support* for the filter. Let the region of support for a filter be denoted by R. Then letting

$$h(m,n)=0\,, \quad (m,n)\notin R\,, \tag{2.2}$$

(2.1) can be written as

$$y(m,n)=\sum_{k=-\infty}^{\infty}\sum_{l=-\infty}^{\infty}h(k,l)x(m-k,n-l)\,. \tag{2.3}$$

In this form the array y is recognized as the 2-D linear convolution of the input array with the array h. This latter array can be seen to be the system output when the input is the 2-D unit impulse.

$$\delta(m,n)=\begin{cases}1, & m=n=0\\ 0, & \text{otherwise}\,.\end{cases} \tag{2.4}$$

The array h is thus called the *impulse response* of the filter. Nonrecursive digital filters are also referred to as finite area impulse response (FIR) filters, since the region of support of the impulse response of these filters is contained in a finite area.

If the input to a 2-D linear, shift-invariant filter is a complex sinusoid of the form

$$x(m,n)=\exp[\mathrm{j}(m\mu+n\nu)] \tag{2.5}$$

the output will be a sinusoid at the same frequencies

$$y(m,n)=H(\mu,\nu)\exp[\mathrm{j}(m\mu+n\nu)]\,. \tag{2.6}$$

The ratio between the output and input sinusoids, $H(\mu,\nu)$, is called the *frequency response* of the filter. It is the Fourier transform of the impulse response. Thus

$$H(\mu,\nu)=\sum_{m=-\infty}^{\infty}\sum_{n=-\infty}^{\infty}h(m,n)\mathrm{e}^{-\mathrm{j}(m\mu+n\nu)}\,, \tag{2.7}$$

$$h(m,n)=\frac{1}{4\pi^2}\int_{-\pi}^{\pi}\int_{-\pi}^{\pi}H(\mu,\nu)\mathrm{e}^{\mathrm{j}(m\mu+n\nu)}d\mu d\nu\,. \tag{2.8}$$

Since h can be computed from H and vice versa, either can be used to completely specify the 2-D filter. As an aside, it should be noted that $H(\mu,\nu)$ is a periodic function of both μ and ν with a period of 2π in each variable.

Two additional observations concerning the frequency responses of 2-D nonrecursive filters can also be made at this point. First, due to the finite region of support of the impulse response h, $H(\mu, v)$ is a two-variable polynomial in the variables $\exp(-j\mu)$ and $\exp(-jv)$. This fact is critical to the design problem. Secondly, the frequency response $H(\mu, v)$ will be a purely real function if and only if

$$h(-m, -n) = h^*(m, n) . \tag{2.9}$$

Filters which have this property are called *zero-phase filters*. From (2.9) it is seen that the region of support of a zero-phase filter is symmetric with respect to rotations of 180° about the origin of the (m, n) plane. If the impulse response coefficients of such a filter are also constrained to be real, then it follows that

$$h(m, n) = h(-m, -n) . \tag{2.10}$$

This constraint can be incorporated into most nonrecursive design algorithms and can be exploited in the implementation of the resulting filters to reduce the total amount of computation.

The filter design problem is concerned with the selection of an array of impulse response coefficients $h(m, n)$, consistent with a given region of support R, which will yield a filter which approximates, in some sense, either an ideal impulse response or an ideal frequency response. Unless the ideal itself has support on R, in which case the design problem is trivial, the filter designer has considerable freedom in deciding how the ideal specifications should be approximated. In the bulk of the remainder of this chapter, three different design algorithms will be presented. Although these represent merely a sampling of techniques which have been proposed, it will be seen that how the approximation is performed will affect not only the complexity of the design algorithm, but also the complexity of the filter implementation.

In Chap. 3 the design of recursive filters is presented. These filters are implemented by difference equations and they are characterized by regions of support which generally have infinite areas. These filters can be unstable (see Chap. 3), and unlike nonrecursive filters they cannot be designed to be implementable while having zero phase. They can also be difficult to design. On the other hand, recursive filters require less storage in their implementation than nonrecursive filters and they may require less computation as well for a given degree of approximation. Each class of filters thus has its advantages and its advocates. For most problems both types of filters should be considered and compared.

2.1.2 Implementation of Nonrecursive Filters

If a filter $h(m, n)$ has support over the region $R = \{(m, n)\colon 0 \leqq m \leqq M-1, 0 \leqq n \leqq N-1\}$, then the convolution sum becomes

$$y(m, n) = \sum_{k=0}^{M-1} \sum_{l=0}^{N-1} h(k, l)x(m-k, n-l) . \tag{2.11}$$

This sum represents an algorithm with which to compute the successive samples of the output array. Each sample of y is seen to require MN multiplications and $MN-1$ two-input additions. The required computation per output sample thus depends linearly on the area of the region of support. It can also be seen that if the assumption is made that all of the samples of x are available, the output samples can be computed in any order.

With real, zero-phase filters, some filter coefficients are equal to others. This can be exploited to halve the number of multiplications necessary to implement the filter. If $R=\{(m,n): -M \leqq m \leqq M, \ -N \leqq n \leqq N\}$, the convolution sum becomes

$$y(m,n)=\sum_{k=-M}^{M}\sum_{l=1}^{N} h(k,l)\,[x(m-k,n-l)+x(m+k,n+l)]$$

$$+\sum_{k=1}^{M} h(k,0)\,[x(m-k,n)+x(m+k,n)]+h(0,0)x(m,n)\,. \tag{2.12}$$

Similar savings can be realized if $h(m, n)$ possesses other types of symmetry.

A nonrecursive filter can also be implemented by means of the 2-D discrete Fourier transform (DFT). This is particularly appealing for high-order filters, due to the computational efficiency of the FFT algorithm for evaluating the DFT. The Fourier transforms of the input and output arrays of a filter are related by

$$Y(\mu,\nu)=X(\mu,\nu)H(\mu,\nu)\,. \tag{2.13}$$

The DFT corresponds to samples of the Fourier transform of a finite area array. If $M \times N$ DFTs of the arrays x, h, and y are computed, it follows that

$$Y(k,l)=X(k,l)H(k,l)\,. \tag{2.14}$$

The output of the filter, $y(m,n)$, can thus be computed as the inverse DFT of the product of the DFTs of the input and the impulse response. However, one must be careful! For this procedure to yield correct results, all of the transforms must be of the same size and that size must be sufficient to uniquely specify the output array. Let

$$R_x=\{(m,n):0\leqq m\leqq M_x, 0\leqq n\leqq N_x\}\,, \tag{2.15a}$$

$$R_h=\{(m,n):0\leqq m\leqq M_h, 0\leqq n\leqq N_h\}\,, \tag{2.15b}$$

$$R_y=\{(m,n):0\leqq m\leqq M_y, 0\leqq n\leqq N_y\} \tag{2.15c}$$

denote the regions of support for x, h, and y. From the convolution sum it follows that

$$M_y=M_x+M_h-1\,, \tag{2.16a}$$

$$N_y=N_x+N_h-1\,. \tag{2.16b}$$

For the inverse DFT to yield the correct array $y(m,n)$, M and N must thus be chosen such that $M \geqq M_y$ and $N \geqq N_y$. The arrays x and h which have smaller regions of support can be defined over larger regions of support through the addition of samples with value zero. Other efficient ways of calculating (2.11) are discussed in Chap. 6.

2.2 The Design of Nonrecursive Filters Using Windows

Two-dimensional nonrecursive filters can be designed using windows using a straightforward extension of the 1-D technique. No essential modifications to the method need to be made to accommodate the increase in dimensionality. With the method, an ideal impulse response is multiplied by a window with finite support to give a filter with finite support. The window can also taper the ideal response to achieve some smoothing in the frequency domain. In the discussions that follow $i(m,n)$ and $I(\mu,\nu)$ will be used to denote the impulse response and frequency response of the ideal filter. The filter being designed will be assigned impulse response $h(m,n)$, frequency response $H(\mu,\nu)$, and region of support R.

2.2.1 The Design Procedure

With the window procedure, the filter $h(m,n)$ is determined as the product of the ideal impulse response array and another array $w(m,n)$, known as the *window function*

$$h(m,n)=i(m,n)w(m,n) . \tag{2.17}$$

Since the ideal is generally presumed to have infinite support, if the support of w is limited to R, the support of h will be limited to R also. Thus this technique produces a filter with the required support. Further, if both $i(m,n)$ and $w(m,n)$ satisfy the zero-phase symmetry condition given by (2.10), then $h(m,n)$ will be zero phase. Since h is formed as a spatial domain product, its frequency response can be found as the 2-D convolution of $I(\mu,\nu)$ with $W(\mu,\nu)$. Specifically,

$$H(\mu,\nu)=\frac{1}{4\pi^2}\int_{-\pi}^{\pi}\int_{-\pi}^{\pi} I(\alpha,\beta)W(\mu-\alpha,\nu-\beta)d\alpha d\beta , \tag{2.18}$$

where $W(\mu,\nu)$ is the Fourier transform of the window function. It can thus be seen that the function W serves to smear the ideal frequency response. For H to closely resemble I, W should approximate an impulse in frequency to the degree possible given that the support of w must be limited to the region R.

What makes the method simple is the fact that the window is chosen to depend only on R, not upon the ideal filter i. Thus once a good window has been discovered for a particular region of support, it can be used for the design of any filter with that region of support.

2.2.2 Window Specification

The requirements for a good 2-D window are the same as those demanded for good 1-D windows, apart from the change in dimensionality. Since many good 1-D windows are known (see, e.g. [2.1]), it is reasonable to seek to convert these to 2-D ones. There are two means by which this can be done, depending upon whether the region R is round or rectangular. The first approach was introduced by *Huang* [2.2]. It forms a 2-D window by sampling a circularly rotated 1-D continuous window function. This will be referred to as the *rotated formulation*

$$w_R(m,n)=w(\sqrt{m^2+n^2})\,. \tag{2.19}$$

Although these windows have a circular region of support, the technique could be generalized to allow for other shaped regions, R, by setting

$$w_R(m,n)=w[f(m,n)]\,. \tag{2.20}$$

To the best of the author's knowledge, this has not been done for any other choice of f. The second formulation, the *Cartesian formulation*, forms a 2-D window with a square or rectangular region of support by forming the Cartesian (outer) product of two 1-D windows.

$$w_C(m,n)=w_1(m)w_2(n)\,. \tag{2.21}$$

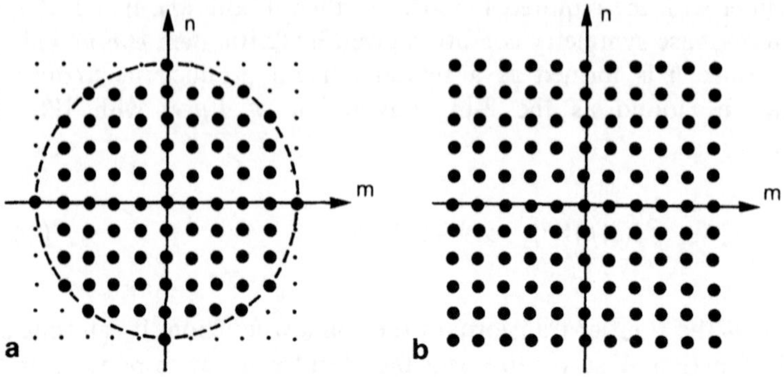

Fig. 2.1. Typical regions of support which could result from the two 2-D window formulations, **(a)** Rotated window; **(b)** Cartesian product window

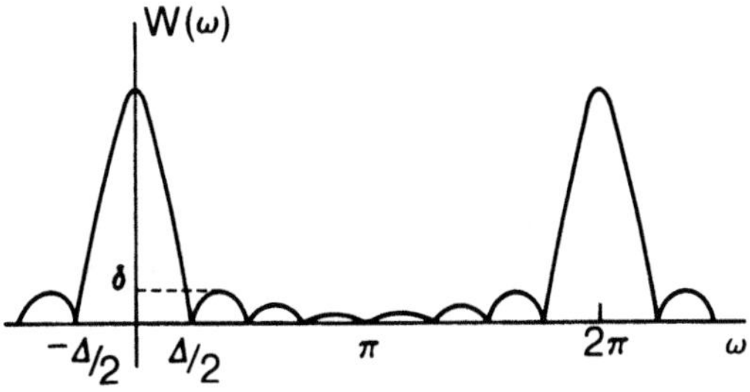

Fig. 2.2. The Fourier transform of a typical one-dimensional window function with main lobe width Δ and sidelobe height δ

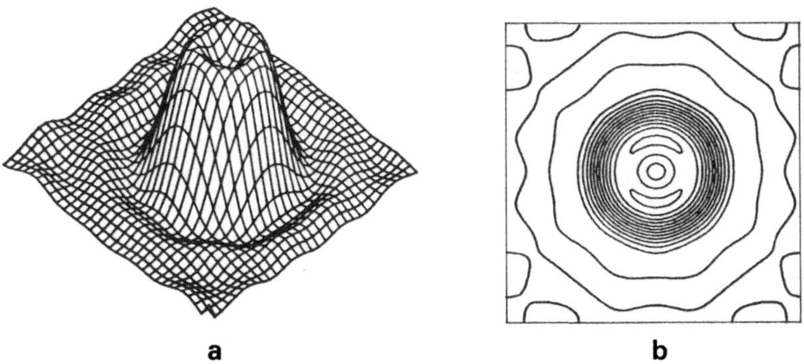

Fig. 2.3a, b. The frequency response of an 11×11 nonrecursive filter designed using a rotated Kaiser window. (**a**) Perspective plot; (**b**) Contour plot

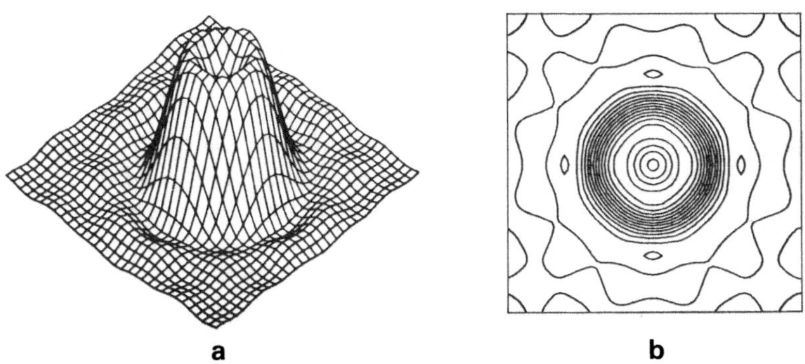

Fig. 2.4a, b. The frequency response of an 11×11 nonrecursive filter designed using an outer product of two 1-D Kaiser windows. (**a**) Perspective plot; (**b**) Contour plot

The regions of support for these two windows are given in Fig. 2.1. Since

$$W_R(\mu, \nu) \simeq W(\sqrt{\mu^2 + \nu^2}), \tag{2.22}$$

$$W_C(\mu, \nu) = W_1(\mu) W_2(\nu) \tag{2.23}$$

it follows that if w, w_1, and w_2 are good 1-D windows, w_R and w_C should be satisfactory 2-D windows.

Figure 2.2 shows the Fourier transform of a typical 1-D window. It is characterized by a large low frequency lobe and it exhibits oscillatory behavior for higher frequencies. The better the window, the smaller the width of the main lobe and the lower the height of the side lobes. For a fixed R, however, there is a necessary tradeoff between these two parameters. A popular choice for the 1-D prototype window to be used with either formulation is the Kaiser window which assumes the form

$$w(t)=\begin{cases}\dfrac{I_0(\alpha\sqrt{1-(t/\tau)^2})}{I_0(\alpha)}, & |t|<\tau \\ 0, & \text{otherwise}\end{cases} \tag{2.24}$$

where $I_0(x)$ is the modified Bessel function of the first kind of order zero. The parameter α is chosen by the designer; it controls the tradeoff between main lobe width and side lobe height.

In Figs. 2.3 and 2.4 are shown the frequency responses of two filters designed using windows. In the first a rotated Kaiser window ($\alpha=0$) was used to design an 11×11 lowpass with a circular cutoff frequency of 0.4π. The second filter is identical except that a Cartesian product window was used. The contour plots show the amplitude of H for contour values between -0.1 and 1.1 in increments of 0.1.

2.3 The Design of Nonrecursive Filters Using Transformations

One of the more popular methods for the design of high-order zero-phase 2-D nonrecursive filters is through a transformation of a 1-D design. This approach is both straightforward to understand and straightforward to apply, rivalling the window method in these respects. Importantly, filters designed by this method can often be implemented with considerably less computational effort than filters designed by the window method. In addition to these features associated with the technique, the filters can often be shown to be optimal in the Chebyshev sense as well.

This design method was introduced by *McClellan* [2.3] in 1972, but it has subsequently been refined and modified. The associated filter implementation was not discovered until three years later [2.4]. The treatment below will include many of the recent refinements to the method [2.5–7].

2.3.1 The Transformation Method

This design method is basically a transformation of a 1-D zero-phase nonrecursive filter into a 2-D zero-phase filter by means of a substitution of variables. The design of a high-order 2-D filter is thus decoupled into the design of a high-order 1-D filter and a low-order transformation. One limitation of the method is that it can only be used in the design of zero-phase filters. This limitation may not be severe since many applications demand the use of zero-phase filters. If zero phase is not desired, then the window method should be used.

The impulse response of a 1-D zero-phase filter is Hermetian symmetric, which implies that it must be of odd length. Let the support of such a filter be over the range $[-N, N]$ and let its frequency response be denoted by $H(\omega)$. One can then write

$$H(\omega)=h(0)+\sum_{n=1}^{N} h(n)(e^{-j\omega n}+e^{j\omega n})$$

$$=\sum_{n=0}^{N} a(n)\cos\omega n\ , \tag{2.25}$$

where

$$a(n)=\begin{cases} h(0), & n=0 \\ 2h(n), & n>0\ . \end{cases} \tag{2.26}$$

Using the Chebyshev polynomials $\{T_n[x]\}$, $\cos\omega n$ can be expressed as a polynomial of degree n in $\cos\omega$. Thus

$$\cos\omega n=T_n[\cos\omega] \tag{2.27}$$

and

$$H(\omega)=\sum_{n=0}^{N} a(n)T_n[\cos\omega]\ . \tag{2.28}$$

At this point a 2-D frequency function, called a *transformation function*, can be substituted for $\cos\omega$. Denoting the transformation function by $F(\mu, v)$, the 2-D frequency response $H(\mu, v)$ results, where

$$H(\mu, v)=\sum_{n=0}^{N} a(n)T_n[F(\mu, v)]\ . \tag{2.29}$$

For any of this manipulation to be meaningful, several things must be shown. First, it must be established that $H(\mu, v)$ is, in fact, the frequency response of a 2-D nonrecursive filter. Second, it must be shown that the nature of H can be predicted from knowledge of $H(\omega)$ and $F(\mu, v)$. Finally a procedure must be presented for choosing a particular $H(\omega)$ and $F(\mu, v)$ to satisfy a given set of frequency specifications on $H(\mu, v)$.

Clearly not all of these can be accomplished without imposing some constraints on the class of allowable transformation functions. For $H(\mu, v)$ to correspond to the frequency response of a nonrecursive filter, it can be shown

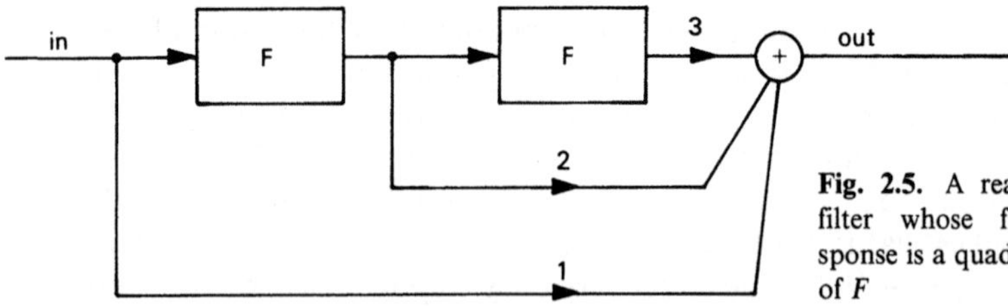

Fig. 2.5. A realization of a filter whose frequency response is a quadratic function of F

that $F(\mu, \nu)$ itself must be the frequency response of a 2-D nonrecursive filter, but F can be of low order. The simplest such choice for $F(\mu, \nu)$, is for it to be the frequency response of a 3×3 filter. In that case

$$F(\mu,\nu)=A+B\cos\mu+C\cos\nu+D\cos(\mu-\nu)+E\cos(\mu+\nu)\,, \tag{2.30}$$

where A, B, C, D, and E are free parameters.

Since $T_n[x]$ is a polynomial of degree n in x, it can be seen that $H(\mu, \nu)$ must also be a polynomial in F but of degree N. The filter with frequency response $H(\mu, \nu)$ could thus be realized through cascade and parallel combinations of identical subnetworks each with frequency response F. Figure 2.5, for example, shows a realization of a filter with the overall frequency response

$$H(\mu,\nu)=3F^2(\mu,\nu)+2F(\mu,\nu)+1\,. \tag{2.31}$$

In the spatial-domain multiplication by F is replaced by convolution by $f(m,n)$, the inverse Fourier transform of F. If $f(m,n)$ possesses a $(2P+1)\times(2P+1)$ point region of support, the N-fold convolution of $f(m,n)$ with itself will possess a $(2NP+1)\times(2NP+1)$ point region of support, as will the filter $h(m,n)$. Thus h is nonrecursive. If $f(m,n)$ is zero phase, F will be real. If further the prototype impulse response coefficients $h(n)$ are real, then from (2.29) H is seen to be zero phase as well.

Consider a locus of points in the (μ, ν) plane such that F is constant on the locus. From (2.29) H is seen to be constant on that locus as well. Thus contours of constant F (henceforth called *isopotentials*) are also isopotentials of H. On each isopotential the value that H assumes will depend on the 1-D prototype filter through the coefficients $\{a(k)\}$. As an example consider the first-order transformation with $A=-1/2$, $B=C=1/2$, $D=E=1/4$, which was originally due to *McClellan* [4.3]. The isopotential contours for this transformation are shown in Fig. 2.6a for values of the 1-D frequency between $0.1\,\pi$ and $0.9\,\pi$ in increments of $0.1\,\pi$. Under that transformation the 1-D prototype of Fig. 2.6b produces the 2-D filter shown in Fig. 2.6c. The same transformation, when used with the bandpass prototype of Fig. 2.6d produces the 2-D bandpass filter shown in part Fig. 2.6e. The shapes of the isopotential contours are controlled by F. The values of the 2-D frequency response on these contours are controlled by both F and the 1-D prototype filter.

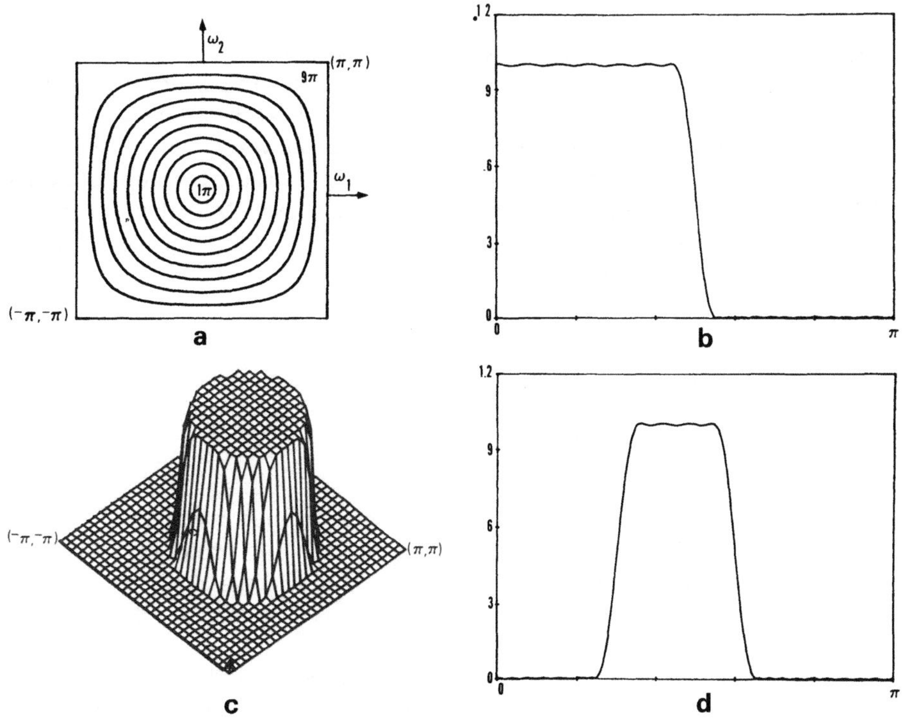

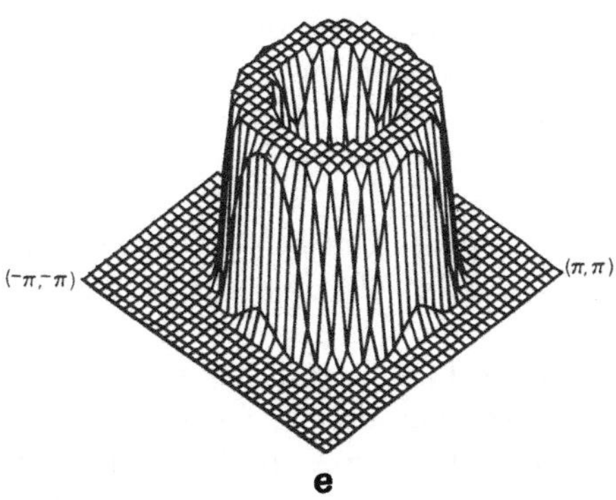

Fig. 2.6a–e. The contours of constant F for a particular first-order transformation function. (**a**) The contours are shown for values of ω in increments of 1π. (**b**) A 1-D prototype lowpass. (**c**) The 2-D lowpass which results under the transformation. (**d**) A 1-D prototype bandpass. (**e**) A 2-D bandpass which results under the transformation

2.3.2 The Design of Transformation Functions

The transformation design procedure produces a high-order 2-D zero-phase filter from a 1-D prototype filter and a transformation function $F(\mu, v)$, which is itself a low-order 2-D nonrecursive filter. It is this decomposition of the design problem which makes this design philosophy so attractive. There remains, however, the issue of how $H(\omega)$ and $F(\mu, v)$ should be chosen to satisfy a given set of 2-D filter specifications. This is the subject of this section. A more complete discussion of this problem is given in [2.5, 8].

It was argued in the last section that isopotential shape depended only on F, whereas the value which $H(\mu, \nu)$ assumed on a particular isopotential depended both upon F and upon the prototype response $H(\omega)$. In performing a design, therefore, $F(\mu, \nu)$ is typically chosen first, then once F has been determined $H(\omega)$ can be chosen. Inasmuch as the selection of $H(\omega)$, however, is the easier of the two problems, it is convenient to consider it first. Consider the transformation function given by (2.30) which was used in Fig. 2.6. Using the values for the free parameters chosen for the figure, by evaluating the 2-D frequency response along the μ axis, one obtains

$$H(\mu, 0) = \sum_{n=0}^{N} a(n) \mathrm{T}_n[F(\mu, 0)]$$

$$= \sum_{n=0}^{N} a(n) \mathrm{T}_n[\cos \mu] = H(\mu) , \qquad (2.32)$$

where the one-dimensional H is the prototype frequency response. A cross-sectional view of the 2-D filter thus looks like the 1-D prototype. If the 2-D filter is to be a lowpass with a cutoff frequency at a radius of $\pi/2$ radians, then $H(\omega)$ should be a 1-D lowpass with a cutoff at $\pi/2$ radians. If $H(\mu, \nu)$ is to be equiripple, then $H(\omega)$ should be equiripple, etc.

For higher-order transformations, the details are only lightly more involved; the approach which should be followed is the same. For each isopotential an ideal value for $H(\mu, \nu)$ can be assigned. Suppose $H(\mu, \nu) = \alpha$ on a particular isopotential. Then $H(\omega)$ should also equal α for

$$\omega = \cos^{-1}[F(\mu, \nu)] . \qquad (2.33)$$

In this fashion the entire ideal prototype response can be constructed. The actual $H(\omega)$ can then be found by any of a number of 1-D nonrecursive filter design methods.

A variety of algorithms have been developed for determining the transformation function $F(\mu, \nu)$ which is best for a particular design. These algorithms fall into two types. In the first group all of the degrees of freedom of the transformation function are used to optimize contour shape. In the second type of algorithm, the shape is controlled only indirectly; $F(\mu, \nu)$ is chosen to optimize the performance of the overall filter. Since the contour matching algorithms are more straightforward, it is perhaps appropriate to look at one of them first.

The method will be described by means of the following example. Assume that a lowpass filter is desired with an elliptic passband. The passband edge should then approximate the contour

$$\left(\frac{\mu}{\mu_0}\right)^2 + \left(\frac{\nu}{\nu_0}\right)^2 = 1 . \qquad (2.34)$$

A first-order transformation will be used. Since this filter is even with respect to both the μ and v variables, the transformation function can assume the restricted form

$$\cos\omega = F(\mu, v) = A + B\cos\mu + C\cos v + D\cos\mu\cos v\ . \tag{2.35}$$

This equation can be solved for, say v, as a function of the other variables, yielding

$$\begin{aligned} v &= \cos^{-1}\left[\frac{\cos\omega - A - B\cos\mu}{C + D\cos\mu}\right] \\ &= G(\mu, \omega, A, B, C, D)\ . \end{aligned} \tag{2.36}$$

An error function can then be defined

$$E_1(\mu) = G(\mu, \omega_0, A, B, C, D) - v_0\sqrt{1 - \left(\frac{\mu}{\mu_0}\right)^2}\ , \tag{2.37}$$

which measures the difference between the isopotential at frequency ω_0 and the desired band edge. This error is a nonlinear function of the five parameters ω_0, A, B, C, and D, which can be selected to minimize some function of $E_1(\mu)$, such as perhaps its maximum absolute value, or its integral squared value for μ in the range $[-\mu_0, \mu_0]$.

There are two difficulties with this approach. First the error is a nonlinear function of its parameters, necessitating the use of nonlinear optimization routines. Secondly, for transformations of higher than first order, an explicit expression for G cannot be found. A minor reformulation of this procedure results in a linear approximation.

If the transformations were exact than as the elliptical contour was transversed, F would remain constant. If the mapping is not exact, however, F will vary. Consider the variation in F to be an error, $E_2(\mu)$, to be minimized. Then

$$E_2(\mu) = \cos\omega_0 - F(\mu, v_0\sqrt{1 - (\mu/\mu_0)^2})\ . \tag{2.38}$$

The error $E_2(\mu)$ is a linear function of A, B, C, D, and $\cos\omega_0$. Thus either linear programming can be used to determine the parameters which minimize the maximum absolute value of E_2 or least squares methods could be used.

Since $F(\mu, v)$ replaces $\cos\omega$ in the 1-D prototype frequency response, the transformation function should satisfy the relation

$$-1 \leqq F(\mu, v) \leqq 1\,, \quad 0 \leqq \mu \leqq \pi\,, \quad 0 \leqq v \leqq \pi\ . \tag{2.39}$$

In the procedure described above this constraint has not been imposed. Fortunately, the transformation function can be chosen without regard to these bounds. A simple linear transformation on $F(\mu, \nu)$ of the form

$$F'(\mu, \nu) = C_1 F(\mu, \nu) - C_2 , \quad C_1 \neq 0 \tag{2.40}$$

can later be applied. The isopotentials of F' possess the same shape as the isopotentials of F. If F_{max} and F_{min} are the maximum and minimum values of $F(\mu, \nu)$, then if C_1 and C_2 are chosen according to

$$C_1 = \frac{2}{F_{max} - F_{min}} , \tag{2.41a}$$

$$C_2 = C_1 F_{max} - 1 , \tag{2.41b}$$

F' will both be appropriately bounded and possess isopotentials of the desired shape. The value of ω which is associated with each isopotential is also changed by this linear transformation.

This procedure can be straightforwardly extended to allow for higher-order transformations or to allow for the shapes of more than one contour to be simultaneously optimized. The interested reader is referred to either [2.5] or [2.8].

The above algorithm is useful for transformation functions of any degree. For transformations of degree greater than one, however, it is possible to design a transformation function which simultaneously defines isopotential shape and minimizes the transition width of the final 2-D filter.

The design problem for 2-D equiripple filters is normally stated in the following terms (see Sect. 2.4): find the $(2N+1) \times (2N+1)$ nonrecursive filter which best approximates, in the Chebyshev sense, an ideal frequency response which has value 1 in its passband $\{(\mu, \nu) \in I_p\}$ and value 0 in its stopband $\{(\mu, \nu) \in I_s\}$. The best approximation, which is not necessarily unique, is one for which the maximum absolute value of the (possibly weighted) error function on $[(\mu, \nu) \in I_p \cup I_s]$ is a minimum.

The transformation design problem can be stated in much the same terms. Given regions I_p and I_s, what is the best transformation of degree $P \times Q$ in the sense that the deviation between the 2-D filter which ultimately results and the ideal filter response is minimized over the union of the passband(s) and stopband(s)? The error magnitude of $H(\mu, \nu)$ will effectively be minimized when the transition bandwidth of the 1-D prototype is maximized. The problem can be rephrased mathematically in the following fashion. Find an F such that

$$\begin{aligned} &\cos \omega_s \geqq F(\mu, \nu) \geqq -1 , \quad && (\mu, \nu) \in I_s \\ &\cos \omega_p \leqq F(\mu, \nu) \leqq 1 , \quad && (\mu, \nu) \in I_p . \end{aligned} \tag{2.42}$$

The coefficients of F are chosen to maximize $\omega_s - \omega_p$. This is a nonlinear optimization problem. It can be linearized by maximizing instead the quantity $\cos \omega_p - \cos \omega_s$. The latter optimization can be solved as a linear program. The values $\cos\omega_p$ and $\cos\omega_s$ are the cosines of the 1-D passband and stopband cutoff frequencies. They are then used to design the 1-D prototype. The wider the transition band, $\omega_s - \omega_p$, of the prototype the smaller the passband and stopband ripple heights in the prototype. These ripple heights then become the 2-D ripple heights because of the nature of the transformation method.

This formulation of the problem tends to give better transformation functions than the contour matching formulation, but for first-order transformations the two approaches are roughly comparable. This approach does require a linear program for its solution, however, and it does not work well on other than piecewise constant ideal functions.

Because the transformation design problem is so much like a nonrecursive filter design problem, it is possible to derive transformation functions from low-order nonrecursive filters. This is especially promising if low-order equiripple designs can be obtained.

Suppose a $(2P+1)\times(2Q+1)$ point 2-D equiripple filter with passband region I_p and stopband region I_s is available. Such a filter, for example, might result from the design algorithm presented in Sect. 2.4. If δ_p and δ_s are maximum errors of that filter in the passband and stopband and if $\hat{F}(\mu, v)$ is its frequency response, a transformation function can be obtained as

$$F(\mu, v) = C_1 \hat{F}(\mu, v) - C_2 \tag{2.43}$$

with

$$C_1 = \frac{2}{1+\delta_p+\delta_s}, \tag{2.44}$$

$$C_2 = \frac{-\delta_s+\delta_p+1}{1+\delta_p+\delta_s}. \tag{2.45}$$

The 1-D prototype filter should be designed to be equiripple (designed perhaps with the *Parks–McClellan* algorithm [2.16]) with passband and stopband cutoff frequencies defined by

$$\omega_p = \cos^{-1}\frac{1+\delta_p-3\delta_p}{1+\delta_p+\delta_s}, \tag{2.46}$$

$$\omega_s = \cos^{-1}\frac{3\delta_s-\delta_p-1}{1+\delta_p+\delta_s}. \tag{2.47}$$

The passband and stopband of $H(\mu, v)$, the final 2-D filter, will be I_p and I_s, and the passband and stopband deviations will be identical to the passband and stopband deviations of the prototype. The filter will have slightly greater deviation than one obtained from (2.42), but the small sacrifice is well worth the computational savings.

2.3.3 Implementation of Transformed Filters

The nonrecursive filters which result from the transformation method, like any nonrecursive filters, can be implemented directly using the convolution sum or they can be implemented using the DFT. In addition, for these filters there is a third implementation which exploits the structure inherent in their design. For filters of moderate order, this third implementation can be more efficient than either convolution or the DFT. This special implementation, in itself, may be sufficient motivation for the use of transformation designs in some applications.

The key to the implementation is (2.29) which is reproduced below

$$H(\mu,\nu)=\sum_{n=0}^{N} a(n)\mathrm{T}_n[F(\mu,\nu)]\ .$$

The special implementation is a digital network which will realize this frequency response. It makes use of a recurrence formula for the Chebyshev polynomials

$$\mathrm{T}_0[x]=1\ , \tag{2.48a}$$

$$\mathrm{T}_1[x]=x\ , \tag{2.48b}$$

$$\mathrm{T}_n[x]=2x\mathrm{T}_{n-1}[x]-\mathrm{T}_{n-2}[x]\ . \tag{2.48c}$$

If x is replaced by $F(\mu,\nu)$, the recursion becomes

$$\mathrm{T}_n[F(\mu,\nu)]=2F(\mu,\nu)\mathrm{T}_{n-1}[F(\mu,\nu)]-\mathrm{T}_{n-2}[F(\mu,\nu)]\ . \tag{2.49}$$

Using this recursion, a signal with Fourier transform $\mathrm{T}_n[F(\mu,\nu)]$ can be produced from one with Fourier transform $\mathrm{T}_{n-1}[F(\mu,\nu)]$ and one with Fourier transform $\mathrm{T}_{n-2}[F(\mu,\nu)]$ by means of the digital network shown in Fig. 2.7. Recognizing now that $\mathrm{T}_{n-1}[F(\mu,\nu)]$ is available from two lower-order polynomials, a ladder network with N outputs can be constructed, where the frequency response between the input and the nth output is $\mathrm{T}_n[F(\mu,\nu)]$. Equations (2.48a, b) can be used to initiate the ladder. If these outputs are

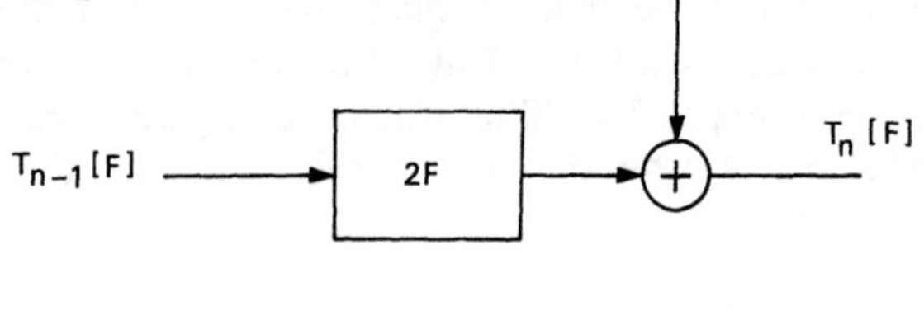

Fig. 2.7. A realization of the Chebyshev recursion relationship using a 2-D nonrecursive digital filter

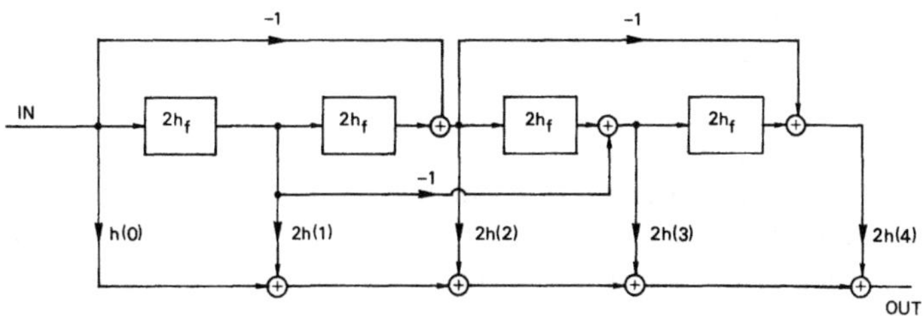

Fig. 2.8. A network realization of a nonrecursive filter designed using a transformation

weighted and summed, according to (2.29), a network with frequency response $H(\mu, \nu)$ results. The network which accomplishes this is shown in Fig. 2.8.

Each of the subnetworks in that figure corresponds to a $(2P+1)\times(2Q+1)$ point 2-D zero-phase nonrecursive digital filter with frequency response $F(\mu, \nu)$ [or $2F(\mu, \nu)$]. For the first-order transformation given by

$$F(\mu, \nu)=A+B\cos\mu+C\cos\nu+D\cos(\mu-\nu)+E\cos(\mu+\nu) \tag{2.50}$$

these filters have a (3×3) point impulse response array $h_f(m,n)$, defined by

$$h_f(m,n)=\begin{cases} A, & m=n=0 \\ B/2, & m=\pm 1, n=0 \\ C/2, & m=0, n=\pm 1 \\ D/2, & m=\pm 1, n=m \\ E/2, & m=\pm 1, n=-m\,. \end{cases} \tag{2.51}$$

Each of these filters can be realized using 5 multiplications per output sample and eight additions. Since there are N of these in the structure, the total computation required for the realization is $6N+1$ multiplications and $10N-2$ additions per output sample. Additionally each filter h_f requires storage of four rows of data. Thus, a total of $4N$ rows of storage are required.

An important feature of the structure is its modularity. To change the 1-D prototype filter, one merely changes the weighting coefficients $\{h(n)\}$. To change the transformation function, all that must be changed are the coefficients within the subfilters labeled h_f.

This same structure can be used for filters of any dimensionality. To implement an N-dimensional filter designed by means of a transformation, one would simply replace h_f by the impulse response corresponding to the N-dimensional transformation function. The actual impulse response $h(m,n)$ never needs to be computed. *McClellan* and *Chan* [2.6] have shown that filters implemented in this fashion possess good arithmetic quantization properties when the filter must be realized with finite precision arithmetic.

2.3.4 A Comparison of the Transformation Implementation with Other Implementations

In both the number of required arithmetic operations and the required words of storage, this special implementation is intermediate between direct convolution and the FFT implementation of a nonrecursive filter. To make the comparison precise assume that a $(2N+1)\times(2N+1)$ point filter is designed by means of a first-order transformation. If this filter is implemented by means of the convolution sum, $(2N+1)^2$ multiplications and a like number of additions per output point would be required. In addition, $2N+1$ rows of the input array must be stored. By exploiting the fact that the filter is zero phase, the number of multiplications can be halved, but the number of additions and the amount of storage cannot. Additional symmetry in the impulse response coefficients of the filter can similarly be exploited to reduce further the number of multiplications.

The same filter when implemented by means of the transformation implementation requires $6N+1$ multiplications and $10N-2$ additions for each output sample plus $4N$ rows of storage. Thus, with this implementation the complexity is proportional to N rather than to N^2.

Another possibility for implementing these filters is through the use of the DFT implemented using a 2-D fast Fourier transform algorithm. Here the amount of computation depends upon the size of the input array which complicates the comparison. On the average, however, the number of multiplications per output point is proportional to the logarithm base two of the number of samples in the output array if a power-of-two FFT is used. The number of additions is also proportional to this logarithm. This reduction in computation is achieved at the expense of increased storage and secondary data handling (see Chap. 5). If the latter is to be minimized, primary storage must be

Table 2.1 The number of multiplications per output point required to filter a 1024×1024 point input array with a $(2N+1)\times(2N+1)$ point zero-phase filter using direct convolution, the FFT, and the special implementation for transformation designs

N	Conv.	FFT	Spec.
5	36	46	31
10	121	46	61
20	441	46	121
40	1681	46	241

available to hold the entire output array. Because it depends upon so many machine dependent parameters, a further comparison with the FFT implementation will not be performed here, although these calculations are not particularly difficult once the constraints of a particular machine are known. It should be stated, however, that the computational savings realizable with the FFT can be huge if the filter order is large.

In Table 2.1 it is assumed that a 1024×1024 point input array is filtered by a $(2N+1) \times (2N+1)$ point nonrecursive filter designed using a transformation of first order. The table gives the number of multiplications required per output point. It is assumed that the filter possesses no symmetries, other than that implied by the zero-phase condition. Were other symmetries present, the number of multiplications with both direct convolution and the special implementation would be reduced.

The transformation design procedure illustrates the close coupling that can exist between filter design and implementation in the 2-D case.

2.4 The Design of 2-D Equiripple Nonrecursive Filters

A number of algorithms have been developed for the design of two-dimensional zero-phase nonrecursive digital filters which are optimal in the Chebyshev or minimax sense. Compared to the window and transformation methods, these algorithms are all computationally slow. The design times grow with the order of the filter, being roughly proportional to the third power of the number of free parameters.

One approach to the design is to use a straightforward application of linear programming. This was done by *Hu* and *Rabiner* in 1972 [2.9]. The extensive computation of this approach motivated *Fiasconaro* [2.10], in an earlier volume of this series, to develop a more sophisticated adaptive algorithm. Independent applications of the ascent algorithm [2.11] by *Kamp* and *Thiran* [2.12] and *Mersereau* et al. [2.13] led to significant reductions in design time. These techniques are compared and refined in a paper by *Harris* and *Mersereau* [2.14]. The primary differences among all of these algorithms concern their computational speed. All are capable of approximating arbitrary linear-phase frequency responses, but all are limited by computational complexity considerations to the design of filters with fewer than approximately 30 degrees of freedom. This corresponds to a circularly symmetric design with a 15×15 point rectangular region of support.

Although the equiripple design algorithms possess nowhere near the efficiency of the window or transformation procedures, there are three reasons why it is beneficial to consider them. First, these algorithms produce filters which are truly optimal in the Chebyshev sense. This can be important in the design of filters of low degree. It is also useful as a benchmark for examining suboptimal designs. Secondly, this approach is completely general in the sense that it can be used to design filters with any magnitude specification. Finally, these filters can be used for the design of transformation functions to use with the transformation method.

2.4.1 The Equiripple Design Problem

Let $H(\mu, v)$ be the frequency response of a realizable system which approximates an ideal frequency response $I(\mu, v)$ on a compact subset, K, of the 2-D frequency plane. For an ideal filter with piecewise constant frequency bands, such as a lowpass, K might be chosen to be the union of the passbands and stopbands, omitting any transition bands where the approximation is not constrained. For a zero-phase filter with a region of support which is a $(2M+1)$ by $(2N+1)$ point rectangle

$$H(\mu, v)=\sum_{m=-M}^{M} \sum_{n=-N}^{N} h(m, n) e^{-jm\mu} e^{-jnv} . \tag{2.52}$$

The impulse response coefficients $\{h(m, n)\}$ are chosen to minimize the Chebyshev error norm measured between I and H.

$$\|H-I\|_K=\max_{(\mu, v) \in K}|H(\mu, v)-I(\mu, v)| . \tag{2.53}$$

The solution H^* which minimizes this norm may or may not be unique. More generality could be introduced into this formulation through the inclusion of a nonnegative weighting function in the error metric, which although straightforwardly done, is omitted here to minimize confusion.

If eightfold (octal) symmetry of the impulse and frequency responses is desired the frequency response can be rewritten as

$$H(\mu, v)=\sum_{i=1}^{R} a(i) \phi_i(\mu, v) , \tag{2.54a}$$

where

$$i=m(m+1)/2+n+1 , \quad 0 \leqq m, n \leqq N , \quad 0 \leqq n \leqq m , \tag{2.54b}$$

$$\phi_i(\mu, v)=\cos^m \mu \cos^n v+\cos^n \mu \cos^m v , \tag{2.54c}$$

$$R=(N+1)(N+2)/2 . \tag{2.54d}$$

Here H is expressed as a linear combination of R approximating or basis functions $\{\phi_i(\mu, v)\}$ each of which is real and has octal symmetry. The coefficients $\{a(i)\}$ of that expansion are related to the impulse response coefficients of the filter. A procedure for determining the $\{h(m, n)\}$ from the $\{a(i)\}$ is given in [Ref. 2.14, Appendix]. The number R represents the number of degrees of freedom in the approximation. It is important to note that H depends linearly on the parameters of the approximation $\{a(i)\}$.

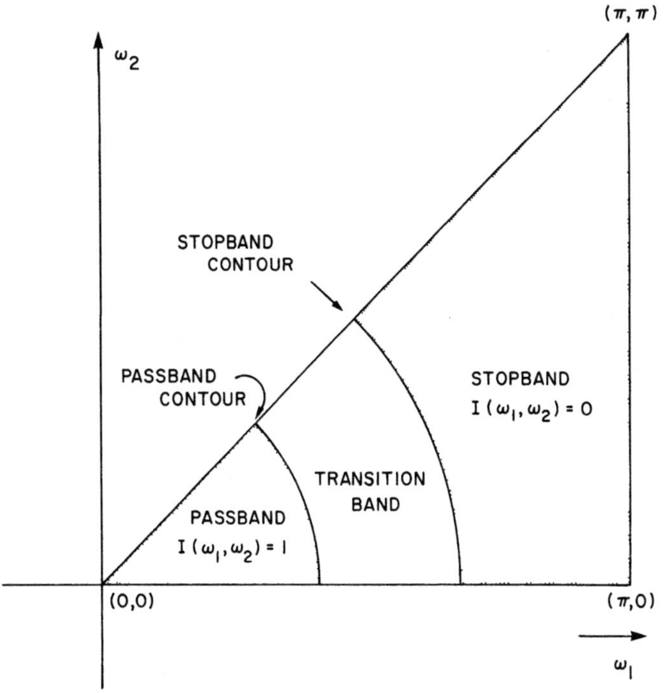

Fig. 2.9. Frequency response specifications of an ideal filter

In addition to reducing the number of degrees of freedom in the approximation, the imposition of octal symmetry allows the error in the frequency response to be minimized only over a single octant of the Fourier plane, as illustrated in Fig. 2.9. This provides obvious computational savings.

The design algorithms which will be presented are iterative and they are guaranteed to converge in a finite number of iterations. In the remainder of this subsection a number of formal definitions will be made and some theorems, which help to identify the best approximation once it has occurred, will also be presented. These are classical results; thus no proofs will be given.

An *extremal point* is a point $x=(\mu, v)$ in the domain of approximation K on which the error function, $|I-H|$ attains its maximum value. The *extremal point set* is the set of all extremal points. The *characteristic vector* associated with the point x is the vector $\Phi(x)=(\phi_1(x), \phi_2(x), \ldots, \phi_R(x))^T$ where the $\{\phi_i(x)\}$ are the basis vectors introduced in (2.54). A point $\boldsymbol{v}$ will be said to lie within the *convex hull* of a finite set of vectors $\{\boldsymbol{v}_i\}$ if and only if there exist constants $\{\alpha_i\}$ which simultaneously satisfy the following three conditions

1) $\boldsymbol{v}=\sum_i \alpha_i \boldsymbol{v}_i$

2) $\alpha_i \geqq 0, \quad \text{for all} \quad i$

3) $\sum_i \alpha_i = 1$.

A *critical point set* will be defined as a set of extremal points of minimum size, such that the zero vector lies within the convex hull of the associated signed characteristic vectors $\sigma(x_i)\Phi(x_i)$ where

$$\sigma(x_i) = \operatorname{sgn}\{H(x_i) - I(x_i)\} . \qquad (2.55)$$

Such a set is of minimum size if the removal of any member would result in the zero vector no longer lying within the convex hull.

Theorem I: A critical point set contains at most $R+1$ members.

Theorem II: $H(x)$ is a best approximation to $I(x)$ in the Chebyshev sense if and only if the zero vector lies in the convex hull of a critical point set associated with the extremal points of the error function.

Theorem III: Let $H(x)$ be a best approximation to $I(x)$ on K. Let C be a critical set associated with the error $H-I$. Then H is a best approximation to I on C.

Theorem IV: Suppose further that C contains p points. Then H is an approximation of maximum norm among all best approximations to I on subsets of K containing p points.

Proofs of these theorems can be found in *Rice* [2.15]. The first theorem limits the dimensionality of the problem. It allows the original approximation problem to be replaced by a search for a critical set of no more than $R+1$ points. The second theorem is a characterization theorem for the best approximation. The remaining two theorems form the basis for the ascent algorithm. That algorithm repeatedly finds a best approximation on sets of p points. Furthermore, successive sets of p points are chosen so that on each the Chebyshev error will be greater than on the previous set. According to the fourth theorem, when this norm attains its maximum value on a set of p points, the best Chebyshev approximation on K has been found and the algorithm terminates. In the next section the detailed operation of the algorithm is presented.

2.4.2 The Ascent Algorithm

In this section a classical algorithm is described for finding an optimal Chebyshev approximation. It is necessary to describe it in some detail since the more efficient algorithms of the following section are refinements on it. The ascent algorithm is a systematic means for finding a critical point set. Once such a set has been found, the determination of the unknown filter coefficients associated with it is straightforward.

The first step is to replace the continuous domain of approximation by a finite set of discrete points K'. Since this discrete set is compact and is a subset of K the four theorems of the previous theorem also are valid on the set K'. For computational convenience, K' is normally taken to consist of samples from K which are regularly spaced in μ and ν. A sample density which will sample the highest frequency basis function ten to twenty times per period in each variable will adequately represent the continuous domain K, provided that additional samples are included along the boundaries of transition regions of the ideal frequency response.

At this point three things need to be done. First a value for p must be chosen. Second a means must be found for finding a best approximation over a set of p points. Finally a means for selecting a new set of p points must be found which will quarantee that the Chebyshev norm ascends.

From the first theorem $p \leqq R+1$. If the characteristic vectors $\Phi(x)$ defined on *any* R points from K' were guaranteed to be linearly independent, $\Phi(x)$ would be said to satisfy the Haar condition and one would set $p=R+1$. With the choice of the $\{\phi_i(x)\}$ given by (2.54c) the Haar condition is not satisfied. In fact it can be shown that no set of 2-D basis functions can satisfy the Haar condition. As a practical matter on most sets of R points from K' the $\Phi(x_i)$ will be linearly independent, thus p will be chosen to be $R+1$. In the special case that a particular set of $R+1$ points from K' is chosen on which the $\Phi(x_i)$ are linearly dependent, the condition is known as a *degeneracy*. A Chebyshev solution cannot be found on a degenerate reference. Thus if a degeneracy occurs, the particular set of p points must be replaced by another. As a practical matter, degeneracies rarely occur.

Next, it is appropriate to consider the problem of finding a best approximation to H on a given set of $R+1$ nondegenerate points from K'. Such a set of $R+1$ points will be called a *reference*. The first step is to establish a linear dependence between the characteristic vectors defined at the $R+1$ points of the reference.

$$\Phi(x_{R+1}) = \sum_{i=1}^{R} \theta_i \Phi(x_i) . \tag{2.56}$$

This is always possible for nontrivial $\{\theta_i\}$ since, by assumption, R of the characteristic vectors are independent and thus form a basis for the space. Letting $\theta_{R+1} = -1$,

$$\sum_{i=1}^{R+1} \theta_i \boldsymbol{\Phi}(x_i) = \mathbf{0} . \tag{2.57}$$

This establishes that the zero vector lies within the convex hull of the characteristic vectors $\{\Phi(x_i)\}$. Thus if the magnitude of the deviation of the error were the same on each point, x_i, the reference would be a critical point set. If d denotes the deviation, the reference corresponds to a critical point set if

$$I(x_i) - \sum_{j=1}^{R} a(j)\phi_j(x_i) = \sigma_i d , \quad i=1,\dots,R+1 . \tag{2.58}$$

The numbers σ_i which are all ± 1 were defined by (2.55). Once these are known, (2.58) becomes a set of $R+1$ linear equations in the $R+1$ unknowns consisting of the coefficients $\{a(j)\}$ and the deviation d. These equations will possess a unique solution if no degeneracy is present. Should one be present, the $\{x_i\}$ can be perturbed slightly.

If (2.58) is first solved for d, the result is

$$d = \frac{\sum_{i=1}^{R+1} \theta_i I(x_i)}{\sum_{i=1}^{R+1} \sigma_i \theta_i} . \tag{2.59}$$

It can be noted parenthetically that the choice $\sigma_i = \operatorname{sgn}(\theta_i)$ maximizes the denominator of this expression and thus of all possible choices of signs minimizes d. It is computationally convenient to use (2.59) to solve for d before solving for the coefficients $\{a(j)\}$. In this way the number of linear equations which must be solved is reduced to R.

To find a set of $R+1$ critical points over the entire domain K' a sequence of best approximations on sets of $R+1$ points are performed such that the deviation on each approximation is higher than on the previous approximation. From the fourth theorem the solution with maximum deviation on a set of $R+1$ points is the optimal solution over K'.

On a typical iteration a best approximation is found on the current reference using the procedure outlined above. Using the coefficients $\{a(j)\}$ thus obtained the error function, $|H-I|$ is then evaluated at all points of K' and the point of K' of maximum error is determined. If that point is already one of the $R+1$ points in the reference, the best approximation on K' has been found and the algorithm terminates. Otherwise, this point is exchanged for a point in the reference to provide the reference for the next iteration. The point which is discarded from the old reference is chosen so that zero will still lie within the convex hull of the $\{\sigma_i \Phi(x_i)\}$ of the new reference. A particularly readable proof that this method of exchange leads to strictly increasing deviations may be found in [2.11].

The following procedure will implement this exchange rule. Let x_α denote the point of K' where, on the current iteration, the error is the largest. Similarly let x_β denote that point of the current reference which will be replaced by x_α in the next reference. Let $\mu = \operatorname{sgn}[H(x_\alpha) - I(x_\alpha)]$ where $H(x)$ is the computed best approximation on that iteration. Now a set of numbers $\{\lambda_i\}$ are determined such that the characteristic vector associated with x_α is a linear combination of the characteristic vectors defined on the reference

$$\mu \Phi(x_\alpha) = \sum_{i=1}^{R+1} \lambda_i \sigma_i \Phi(x_i) . \tag{2.60}$$

Since only R of the characteristic vectors in (2.60) are linearly independent, the term on the right side of the equation due to x_β can be eliminated. From (2.57)

$$\Phi(x_\beta) = \sum_{\substack{i=1\\ i\neq\beta}}^{R+1} \left(-\frac{\theta_i}{\theta_\beta}\right) \Phi(x_i)\,. \tag{2.61}$$

Inserting this result into (2.60)

$$\mu\Phi(x_\alpha) = \sum_{\substack{i=1\\ i\neq\beta}}^{R+1} \left(\lambda_i - \lambda_\beta\sigma_\beta \frac{\theta_i}{\theta_\beta}\,\sigma_i\right) \sigma_i\Phi(x_i)\,. \tag{2.62}$$

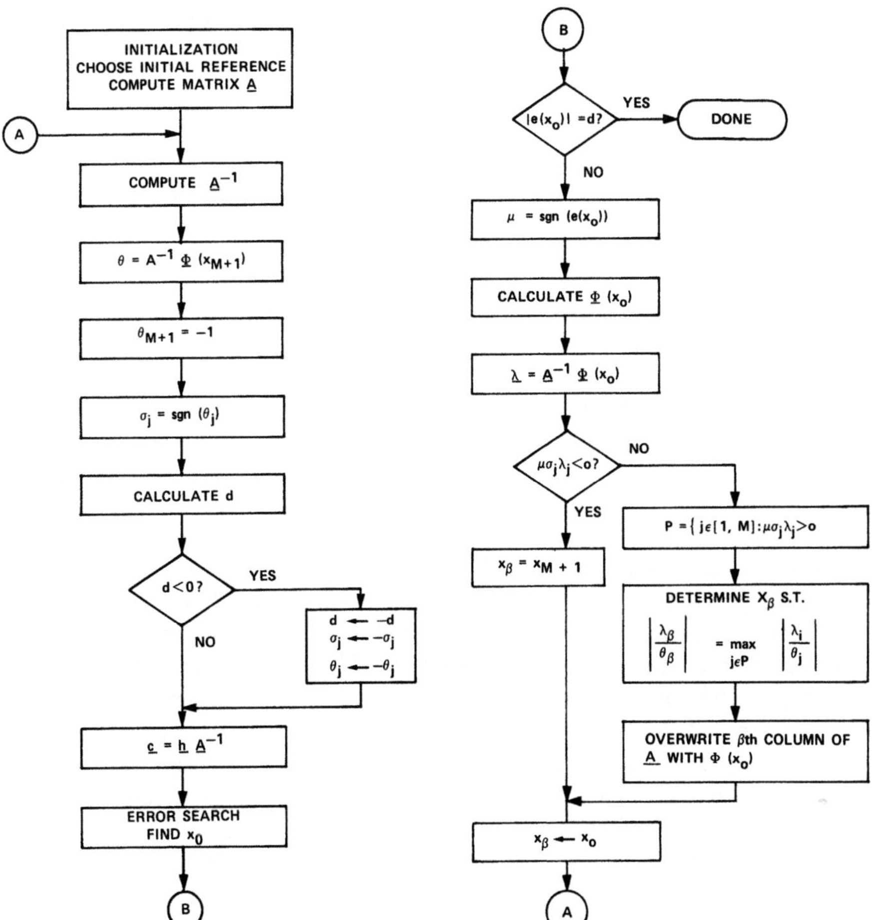

Fig. 2.10. A flow chart of the single exchange ascent algorithm

This expression establishes that the zero vector lies within the convex hull of the vectors of the old reference with x_β replaced by x_α provided that

$$\lambda_i - \lambda_\beta \sigma_\beta \frac{\theta_i}{\theta_\beta} \sigma_i < 0 \qquad i=1,\ldots,R+1\ . \quad i \neq \beta \tag{2.63}$$

Since $\sigma_i \theta_i > 0$ this condition will always be true if β is chosen such that

$$\frac{\lambda_\beta \sigma_\beta}{\theta_\beta} > \frac{\lambda_i \sigma_i}{\theta_i} \qquad i=1,\ldots,R+1\ . \quad i \neq \beta \tag{2.64}$$

This algorithm, the single exchange ascent algorithm, is summarized in the flowchart, which is Fig. 2.10. It requires a large number of iterations, each of which requires that the error function be evaluated over all of K'. This is, by far, the most time-consuming part of the algorithm. Some of the variations on this algorithm such as those of *Kamp* and *Thiran* [2.12] and *Mersereau* et al. [2.13] attempt to speed the algorithm up by reducing the number of error searches over K'. These variations are the subject of the next section.

2.4.3 Variations on the Ascent Algorithm to Improve the Rate of Convergence

Kamp and *Thiran* modified the basic ascent algorithm by replacing the search over K' for a single error maximum on each iteration by a more elaborate search which additionally finds several local maxima. These points are stored in an ordered list such that the first item in the list has the largest error, the second item has the second largest error magnitude, etc. As with the single exchange algorithm, if the item at the head of the list is a point in the reference, the best approximation has been found and the algorithm terminates. The local maxima from the list are used to sequentially replace points in the reference beginning at the top of the list. The exchange rule used is the same one which was used for the single exchange algorithm with the added provision that an exchange is not performed if doing so would displace from the reference one of the local maxima previously inserted during that iteration. A new best approximation is not computed after each exchange; this is only done once per iteration, where an iteration is identified with an error search. Although there is no guarantee that the deviation will increase with every exchange, no failure to converge has been observed that could be attributed to this multiple exchange strategy.

The algorithm of *Harris* and *Mersereau* [2.14] tries to improve upon the convergence rate of the single exchange ascent algorithm through the use of an intermediate sparce grid, G, which, for the lack of a better term, will be termed a *superreference*. The superreference contains more than $R+1$ points but contains considerably fewer points than K'! The single exchange ascent algorithm is used to find a best approximation on G. This can be done reasonably quickly since error searches are confined to G rather than to K'. Once this is done, an error search is done over all of K' and as with the *Kamp* and *Thiran* algorithm, several local maxima are found. These are inserted into G along with the points from the last reference and possibly a few other points as well. The best approximation is then found on the new G and the process repeats. Convergence is obtained when the worst error on K' is equal to the worst error on G.

As G is made larger, the number of error searches over K' is reduced, but the time required to find a best approximation on G is increased. The algorithm seemed to be most efficient when G was chosen to contain approximately $R+10$ points. On each iteration the local maxima from K' were exchanged for these points in G on which the error function was smallest. It was anticipated that this algorithm would converge more rapidly than the algorithm of *Kamp* and *Thiran*, since G retains information about more important points than can be contained in the reference. The expectation was that these points would become local maxima in the error function more than once as the algorithm progressed. Retaining them in the superreference would remove the necessity of rediscovering them during successive error searches. The results of direct comparisons of the two algorithms [2.14] would seem to justify these expectations.

A third refinement of the single exchange ascent algorithm is due to *Harris* [2.14, 17], which used an exchange rule suggested by *Rice* [2.15]. It is similar to the *Harris–Mersereau* algorithm except that a different exchange rule is used within the ascent algorithm. The idea here was that the worst error point on G, x_α, should be exchanged for that point in the reference, x_β, which would maximize the deviation of the best approximation on the new reference. Essentially, the rate of ascent should be maximized. Since x_α can be exchanged for any of the $R+1$ points in the current reference, this approach entails the calculation of the deviation for each of $R+1$ potential new references and choosing the one with the largest deviation.

The calculation of the deviations is facilitated by [2.59]. If the primed quantities denote the values of parameters after the exchange and the unprimed quantities denote their values before the exchange, then the deviation after the exchange can be denoted by

$$d' = \frac{\sum_{j=1}^{R+1} \theta'_j I(x'_j)}{\sum_{j=1}^{R+1} \theta'_j \sigma'_j}. \tag{2.65}$$

The new set of characteristic coefficients $\{\theta'_j\}$ can be computed from the current set $\{\theta_j\}$ by means of the relations

$$\theta'_j = \theta_j - \frac{\theta_\beta}{\lambda_\beta}\lambda_j, \quad j=1,\dots,R+1, \quad j \neq \beta \tag{2.66a}$$

$$\theta'_\beta = \frac{\theta_\beta}{\lambda_\beta} \tag{2.66b}$$

if $\beta \neq R+1$ and

$$\theta'_j = \lambda_j, \quad j=1,\dots,R, \tag{2.66c}$$

$$\theta'_{R+1} = -1 \tag{2.66d}$$

if $\beta = R+1$. The coefficients $\{\lambda_j\}$ were defined earlier.

Although this exchange rule is more complicated than the one derived in the previous section, this algorithm generally converged faster than the other two algorithms, particularly for higher-order filters. This appears to be due to the fact that the total number of exchanges which need to be performed is reduced.

2.4.4 Computational Issues

Of the three algorithms described in the previous section, that due to Harris converges the most quickly. All three algorithms, however, occasionally exhibit numerical difficulties when implemented on computers which perform finite precision arithmetic. In those cases the reference deviation may not ascend from one iteration to the next, references may repeat (which causes the algorithm to cycle forever), or degenerate references can occur. If the algorithm is close to convergence, the filter based on the last reference may be adequate. Otherwise, the algorithm can often be restarted with an approximation to the last reference. In any case, however, these conditions must be identified before any corrective action can be taken. Thus, it is important that the set of reference deviations be monitored by the program implementing the design algorithm. The reasons why these difficulties occur are not completely understood. While it is simple to pin the blame on specific procedures, such as the solution of large sets of linear equations which may be poorly conditioned, the difficulty is probably more subtle. Attempts to use more accurate algorithms for solving linear equations, for example, made these problems worse, not better.

In Fig. 2.11 the frequency response of an 11×11 point octally symmetric lowpass filter is shown which was designed using Harris' algorithm.

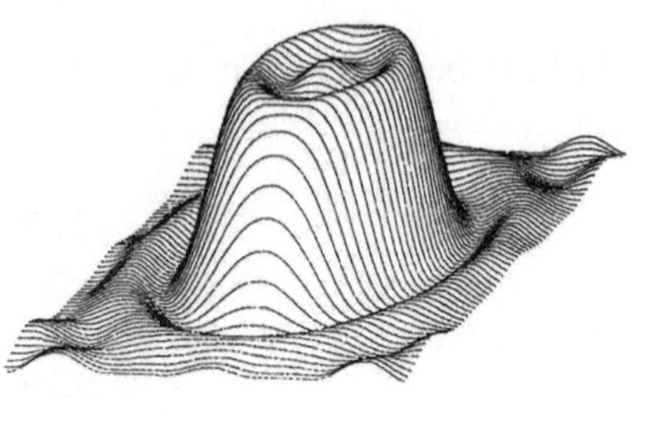

Fig. 2.11. The frequency response of an 11×11 point octally symmetric low-pass nonrecursive filter designed using the exchange algorithm in conjunction with a dynamic spare grid G. The passband and stopband cutoff frequencies are 0.04π and 0.6π and the maximum error in the frequency response is 0.0569

2.4.5 Conditions for Optimality of a Transformed Design

In certain cases nonrecursive designs made by the transformation method are also optimal in the Chebyshev sense. This possibility was first noted by *McClellan* [2.3]. A transformation design can only be optimal in the Chebyshev sense, however, if the passband and stopband contours of the ideal specifications are isopotentials of the transformation, the transformation is of first order, and the 1-D prototype filter is itself optimal in the Chebyshev sense. If these conditions are valid, *Harris* [2.17] has proven the following characterization theorem.

Theorem. A first-order transformation satisfying the conditions mentioned above will produce optimal Chebyshev filters from every optimal prototype if and only if an edge of the square $[-\pi,\pi]\times[-\pi,\pi]$ (the domain of the transformation) maps one-to-one onto the range $[-\pi,\pi]$. One transformation which satisfies this test is the transformation presented in Fig. 2.6.

2.5 Overview

Three methods have been presented for the design of 2-D nonrecursive filters – windows, transformations, and equiripple algorithms. Each has its own advantages. The window method can be used for other than zero-phase filters, and it is easy to use. The transformation method is slightly more complex and can only design zero-phase filters, but under some conditions the resulting filters can be optimal in the Chebyshev sense, and the filters possess an extremely efficient implementation. The equiripple algorithms are by far the most complicated, but they are always optimal. The equiripple designs can also be used as a basis for the design of transformation functions. There are numerous other methods as well, each of which has its own special advantages and drawbacks. These have not been presented here for reasons of space, not because other methods do not possess merit. Also, frankly, the methods described here are those with which the author is most familiar. These three methods, however, do illustrate very different approaches to the filter design problem, and they illustrate the challenge and possibilities of many multidimensional problems.

Acknowledgement. The author has benefitted greatly by being able to work with several gifted graduate students. Many of the results reported here are due to two of these – David B. Harris and Thomas F. Quatieri, Jr. In addition many of the results related to transformation design resulted from a fruitful and pleasant collaboration of the author with Dr. Wolfgang F. G. Mecklenbräuker of Phillips Laboratories.

This was written under a contract from the U.S. Army Research Office, whose continued support is gratefully acknowledged.

References

2.1 F.J.Harris: Proc. IEEE **66**, 51–83 (1978)
2.2 T.S.Huang: IEEE Trans. AU-**20**, 88–90 (1972)
2.3 J.H.McClellan: "The Design of Two-Dimensional Digital Filters by Transformations", in Proc. 7th Ann. Princeton Conf. Information Sciences and Systems (1973) pp. 247–251
2.4 W.F.G.Mecklenbräuker, R.M.Mersereau: IEEE Trans. CAS-**23**, 414–422 (1976)
2.5 R.M.Mersereau, W.F.G.Mecklenbräuker, T.F.Quatieri, Jr.: IEEE Trans. CAS-**23**, 405–414 (1976)
2.6 J.H.McClellan, D.S.K.Chan: IEEE Trans. CAS-**24**, 372–378 (1977)
2.7 R.M.Mersereau: IEEE Trans. CAS-**26** (to be published)
2.8 T.F.Quatieri, Jr.: "The Design of 2-D Digital Filters by Generalized McClellan Transformations"; S.M. Thesis, M.I.T. (1975)
2.9 J.V.Hu, L.R.Rabiner: IEEE Trans. AU-**20**, 249–257 (1972)
2.10 J.G.Fiasconaro: "Two-Dimensional Nonrecursive Filters", in *Picture Processing and Digital Filtering*, 2nd ed., ed. by T.S.Huang, Topics in Applied Physics, Vol. 6 (Springer, Berlin, Heidelberg, New York 1975) pp. 69–129
2.11 E.W.Cheney: *Introduction to Approximation Theory* (McGraw-Hill, New York 1966) pp. 45–51
2.12 Y.Kamp, J.P.Thiran: IEEE Trans. CAS-**22**, 208–218 (1975)
2.13 R.M.Mersereau, D.B.Harris, H.S.Hersey: "An Efficient Algorithm for the Design of Equiripple Two-Dimensional FIR Digital Filters", Proc. 1975 Intern. Symp. Circuits Systems (1975) pp. 443–446
2.14 D.B.Harris, R.M.Mersereau: IEEE Trans. ASSP-**25**, 492–500 (1977)
2.15 J.R.Rice: *The Approximation of Functions, Vol. 2, Nonlinear and Multivariate Theory* (Addison-Wesley, Reading, MA 1969)
2.16 T.W.Parks, J.H.McClellan: IEEE Trans. CT-**19**, 189–194 (1972)
2.17 D.B.Harris: "Iterative Procedures for Optimal Chebyshev Design of FIR Digital Filters"; S.M. Thesis, M.I.T. (197)

3. Design of Two-Dimensional Recursive Filters

P. A. Ramamoorthy and L. T. Bruton

With 17 Figures

This chapter deals primarily with the design of two-dimensional digital recursive filters. Specifically, the design of two types of recursive filters; quarter-plane (QP) recursive filters and half-plane (HP) recursive filters. Many of the recent results on their design are presented. The chapter is organized as follows: in Sect. 3.1, we briefly introduce the idea of QP and HP filters. The different types of QP and HP filters, their stability properties and implementation are discussed there. Section 3.2 contains a detailed discussion of two algorithms for the design of QP filters. These two methods are restricted to special types of QP filters. In Sect. 3.3, a design method for the general class of stable QP recursive filters is presented. The design of nonsymmetric half-plane (NSHP – also known as semicausal) recursive filters is considered in detail in Sect. 3.4. A brief summary and conclusions are presented in Sect. 3.5.

3.1 Preliminaries and Definitions

A two-dimensional (2D) discrete signal $x(m,n)$ is a function of two independent integer variables m and n, where $-\infty<m,\ n<\infty$. A 2D discrete system is represented here by a transformation $T[\cdot]$ which transforms an input discrete signal $x(m,n)$ into an output signal $y(m,n)$. We can obtain useful 2D discrete systems by placing constraints on the transformation $T[\cdot]$. If the system equation $y(m,n)=T[x(m,n)]$ is *linear* in $y(m,n)$ and $x(m,n)$ and *shift-invariant*, then the corresponding system is a linear shift-invariant (LSI) system. Such systems may be characterized by their response $h(m,n)$ to the 2D unit impulse input signal

$$\delta(m,n)=\begin{cases}1 & \text{for} \quad m=n=0\\ 0 & \text{otherwise}\end{cases} \tag{3.1}$$

in terms of the 2D convolution sum

$$y(m,n)=\sum_{k=-\infty}^{\infty}\sum_{l=-\infty}^{\infty}h(k,l)x(m-k,n-l)\,. \tag{3.2}$$

Defining the 2D system transfer function[1]

$$H(z_1, z_2) = \sum_{m=-\infty}^{\infty} \sum_{n=-\infty}^{\infty} h(m,n) z_1^{-m} z_2^{-n} \tag{3.3}$$

it may be shown that

$$Y(z_1, z_2) = H(z_1, z_2) X(z_1, z_2), \tag{3.4a}$$

where

$$\left. \begin{aligned} X(z_1, z_2) &= \sum_{m=-\infty}^{\infty} \sum_{n=-\infty}^{\infty} x(m,n) z_1^{-m} z_2^{-n}; \\ Y(z_1, z_2) &= \sum_{m=-\infty}^{\infty} \sum_{n=-\infty}^{\infty} y(m,n) z_1^{-m} z_2^{-n} \end{aligned} \right\}. \tag{3.4b}$$

We are concerned here with the design of stable 2D recursive filters having a transfer function $H(z_1, z_2)$ with prescribed characteristics.

3.1.1 Quarter-Plane Causal Recursive Filters

A 2D LSI filter with an impulse response $h(m,n)$ is termed QP if $h(m,n)$ has support only in the region $\mathcal{R}_{++} \equiv \{m \geqq 0,\ n \geqq 0\}$ (a quarter-plane) or its rotations $\mathcal{R}_{+-} \equiv \{m \geqq 0,\ n \leqq 0\}$, etc. A 2D recursive filter is characterized by a 2D difference equation relating the input and output as given below:

$$y(m,n) = \sum_{k,l \in \mathcal{S}} a(k,l) x(m-k, n-l) - \sum\sum_{k,l \in \mathcal{S}-(0,0)} d(k,l) y(m-k, n-l). \tag{3.5}$$

Here, the region of support $\mathcal{S}$ for the coefficients $a(m,n)$ and $d(m,n)$ is the finite region $\mathcal{Rf}_{++} \equiv \{m_1 \geqq m \geqq 0,\ n_1 \geqq n \geqq 0\}$ or its rotations. A QP recursive filter is called spatially causal if the region of support $\mathcal{S}$ in (3.5) is $\mathcal{Rf}_{++}$. In this case, we use the past output values which lie below and to the left to calculate the present output. The system function of a 2D causal recursive filter will be of the form

$$H(z_1, z_2) \equiv \frac{A(z_1, z_2)}{D(z_1, z_2)} = \frac{\sum\sum_{k,l \in \mathcal{Rf}_{++}} a(k,l) z_1^{-k} z_2^{-l}}{\sum\sum_{k,l \in \mathcal{Rf}_{++}} d(k,l) z_1^{-k} z_2^{-l}} \equiv \sum\sum_{m,n \in \mathcal{R}_{++}} h(m,n) z_1^{-m} z_2^{-n}. \tag{3.6}$$

That is, the impulse response of a causal recursive filter will have support in the first quadrant $\mathcal{R}_{++}$. Therefore, a causal recursive filter is also termed as ++ recursive filter. It should be noted that ++ recursive filters form only a subset of all infinite impulse response (IIR) filters.

---

1 Here z_1 and z_2 are the unit advance operators in the directions of m and n, respectively.

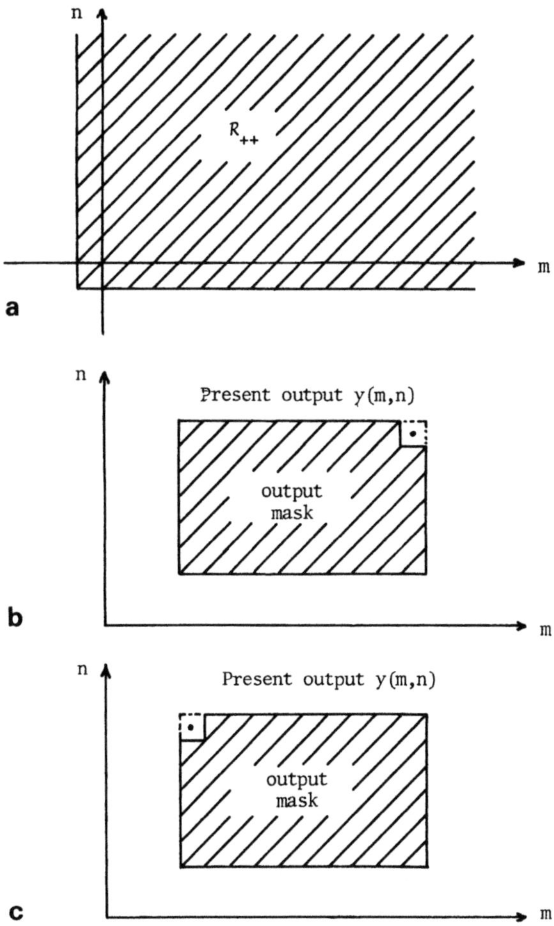

Fig. 3.1. **(a)** The $++$ quarter-plane; **(b)** 2D recursion using a $++$ filter; **(c)** 2D recursion using a $-+$ filter

In a similar manner, by constraining $\mathscr{S}$ in (3.5) to be equal to $\mathscr{R}\!f_{+-} \equiv \{m_1 \geqq m \geqq 0,\ 0 \geqq n \geqq n_1\}$, $\mathscr{R}\!f_{-+}$ and $\mathscr{R}\!f_{--}$, we obtain, respectively, $+-$, $-+$ and $--$ recursive filters. In Fig. 3.1, the QP region $\mathscr{R}_{++}$, a computational sequence using QP causal filter and $-+$ filter are shown.

Special Classes of QP Filters

We can consider restricted regions of a QP that lead directly to special classes of QP filters. For example, if we define $\mathscr{R}^1_{++} \equiv \{m=n=0\} \cup \{m \geqq 0,\ n>0\}$ and change the region of support of the coefficients of the difference equation (3.5) to $\mathscr{R}\!f^1_{++} \equiv \{m=n=0\} \cup \{m_1 \geqq m \geqq 0, n_1 \geqq n \geq 0\}$, we get the special class of causal recursive filters which has a computational sequence as shown in Fig. 3.2a. We can get two more special classes of $++$ recursive filters by defining $\mathscr{R}\!f^2_{++} \equiv \{m=n=0\} \cup \{m_1 \geqq m>0,\ n_1 \geqq n \geqq 0\}$ or $\mathscr{R}\!f^3_{++} \equiv \{m=n=0\} \cup \{m_1 \geqq m$

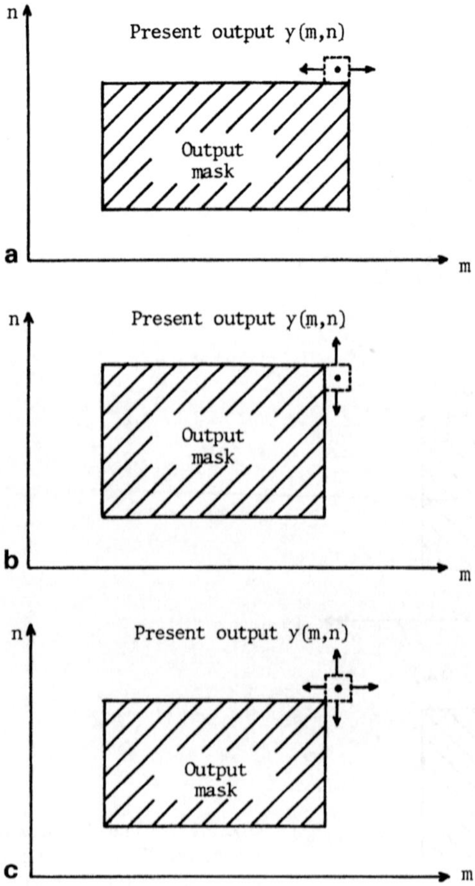

Fig. 3.2a–c. 2D recursion using three special classes of + + filters (the arrows indicate some of the possible directions of recursion to compute the present output)

$>0,\ n_1 \geqq n > 0\}$ (finite regions corresponding to $\mathscr{R}^2_{++}$ and R^3_{++}) and the computational sequences of the output are depicted in Fig. 3.2b, c. In a similar fashion we obtain special classes of QP filters corresponding to $+-$, $-+$ and $--$ recursive filters.

We have many reasons for defining the special classes of QP filters. The one which is obvious from Fig. 3.2 is that the output of 2D QP recursive filters can be calculated using a variety of possible computational sequences which will all result in the same output. Also, these types of filters may be used to understand NSHP recursive filters.

3.1.2 NSHP Recursive Filters

A NSHP is defined in a similar way and is given by the region $\mathscr{R}_{\oplus +} \equiv \mathscr{R}_{++} \cup R^3_{-+} = \{m \geqq 0,\ n \geqq 0\} \cup \{m < 0,\ n > 0\}$, $\mathscr{R}_{\ominus -} \equiv \mathscr{R}_{--} \cup \mathscr{R}^3_{+-}$ (the two signs determine the $\mathscr{R}_{..}$ part and the "$\odot$" determines the $\mathscr{R}^3_{..}$ part by a

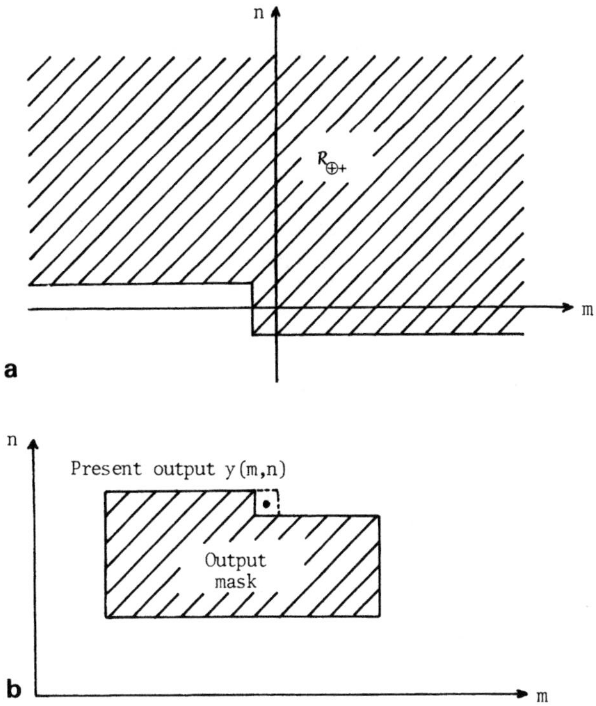

Fig. 3.3. **(a)** $\oplus +$ nonsymmetric half-plane; **(b)** 2D recursion using a $\oplus +$ filter

change of the corresponding sign) or their roations [3.1]. A 2D NSHP filter will have an impulse response which is nonzero only in a particular NSHP. Again, by restricting the numerator and the denominator polynomial of the transfer function to a finite region of a NSHP, we obtain the class of NSHP recursive filters which can be implemented in a recursive fashion. The transfer function of a recursive filter having support in the NSHP $\mathscr{R}_{\oplus +}$ is thus given by

$$H(z_1, z_2) \equiv \frac{A(z_1, z_2)}{D(z_1, z_2)} = \frac{\sum\sum_{\mathscr{R}f_{\oplus +}} a(k, l) z_1^{-k} z_2^{-l}}{\sum\sum_{\mathscr{R}f_{\oplus +}} d(k, l) z_1^{-k} z_2^{-l}} \tag{3.7}$$

and the output $y(m, n)$ is calculated recursively using the 2D difference equation

$$y(m, n) = \sum\sum_{\mathscr{R}f_{\oplus +}} a(k, l) x(m-k, n-l) - \sum\sum_{\mathscr{R}f_{\oplus +} - (0,0)} d(k, l) y(m-k, n-l) . \tag{3.8}$$

By changing the region of support in (3.7, 8) we obtain other NSHP filters. As in the case of QP recursive filters, NSHP recursive filters can be used to realize only a subset of IIR filters. However, because of the ease of implementation (using recursion), QP and NSHP recursive filters form an important class of IIR filters. Figure 3.3 gives the NSHP $\mathscr{R}_{\oplus +}$ and the sequence of computation using a $\oplus +$ NSHP recursive filter. Note from Figs. 3.1, 3 that the main difference between a QP and NSHP filter comes by the way in which the output masks are defined. The output mask for NSHP filters is more general than that for QP filters. Hence we may expect that NSHP filters will be superior to QP recursive filters. We shall go into this point in detail in Sect. 3.4.

3.1.3 Stability of Recursive Filters

To be usable in practical applications, an LSI system has to produce bounded outputs $y(m,n)$ for all bounded inputs $x(m,n)$. This is known as the BIBO stability criterion and is satisfied if and only if

$$\sum\sum |h(m,n)| < \infty\,. \tag{3.9}$$

For the class of filters considered in this chapter (QP and NSHP recursive filters) the impulse response is generally nonzero in an infinite region of m and n, and stability is not always guaranteed. Therefore, it is essential that stability be imposed in any design algorithm. For recursive filters, we can obtain equivalent conditions for stability in terms of the system function, they are given below.

Stability of Causal Recursive Filters

Consider the system function of a causal recursive filter as given in (3.6) with $\mathcal{S} = \mathcal{R}\!f_{++}$. We then have the following stability equivalences:

$$\text{BIBO } \textit{Stability} \underset{\Leftarrow}{\not\Rightarrow} \underbrace{D(z_1,z_2) \neq 0 \text{ in } \bar{U}^2 \equiv \{|z_1| \geqq 1, |z_2| \geqq 1\}}_{\text{min-min phase polynomial in } z_1,\, z_2} \tag{3.10a}$$

$$\text{BIBO } \textit{Stability} \underset{\not\Leftarrow}{\Rightarrow} D(z_1,z_2) \neq 0 \text{ in } \bar{U}^2 - T^2[\equiv (|z_1| = |z_2| = 1)]\,. \tag{3.10b}$$

We now present some examples of 2D QP recursive filter transfer functions to demonstrate the implication of the above equations.
Examples:

$$H_1(z_1,z_2) = \frac{z_1 z_2}{z_1 z_2 + 0.6\, z_1 + 0.8\, z_2 + 0.5} \tag{3.11}$$

represents the transfer function of a stable causal filter since its denominator is a min–min phase polynomial in z_1, z_2.

$$H_2(z_1,z_2) = \frac{(z_1-1)^8\,(z_2-1)^8}{z_1^7 z_2^7 (2 z_1 z_2 - z_1 - z_2)} \tag{3.12}$$

is a stable causal filter transfer function though its denominator (as well as the numerator) has a zero at $z_1 = z_2 = 1$ and is nonzero everywhere else in $\bar{U}^2$ and

$$H_3(z_1,z_2) = \frac{(z_1-1)(z_2-1)}{2 z_1 z_2 - z_1 - z_2} \tag{3.13}$$

represents the transfer function of a causal filter which is unstable. It can be observed that the denominator (and the numerator) of $H_3(z_1,z_2)$ is nonzero in

$\bar{U}^2$ except at $z_1 = z_2 = 1$. For a detailed description of the BIBO stability of causal recursive filters the reader is referred to [3.2] and Chap. 4.

Stability of NSHP Recursive Filters

By considering the system function of a $\oplus +$ NSHP filter as in (3.7), we have the following properties:

$$\text{BIBO } \textit{Stability} \underset{\not\Rightarrow}{\Leftarrow} \begin{cases} \underbrace{D(z_1, z_2) \neq 0 \text{ in } \bar{U}^{20} \equiv \{|z_1| = 1, |z_2| \geqq 1\}} \\ \text{mixed-min phase polynomial in } z_1, z_2 \\ D(z_1, \infty) \neq 0 \text{ for } \{|z_1| \geqq 1\} \end{cases} \tag{3.14a}$$

$$\text{BIBO } \textit{Stability} \underset{\Rightarrow}{\not\Leftarrow} \begin{cases} D(z_1, z_2) \neq 0 & \text{in} \quad \bar{U}^{20} - T^2 \\ D(z_1, \infty) \neq 0 & \text{for} \quad \{|z_1| \geqq 1\}. \end{cases} \tag{3,14b}$$

From the preceding statements and examples, it becomes obvious that if the denominator polynomial is void of zeros in a certain region [(3.10a, 14a)], then the stability is guaranteed. However, there exist transfer functions having denominator polynomials which have zeros on the poly-boundary T^2 and still may represent stable filters. This situation of having nonessential singularities of the second kind [3.3] on T^2 is mostly of theoretical interest. Therefore, hereafter the design of recursive filters will be restricted to stable transfer functions satisfying the conditions in (3.10a) or (3.14a).

3.1.4 Use of Bilinear Transformation in Stability

In the last subsection, we noted that the denominator of the system function of a stable recursive filter has to be nonzero in a certain region of the variables z_1 and z_2. In order to test this condition, we can use a transformation of variables which changes the polydomain in which the system function has to be free of poles to other, more suitable, domains. One such transformation which has been used extensively is the bilinear transformation and is given by

$$z_i = \frac{1 + s_i}{1 - s_i}; \qquad i = 1, 2. \tag{3.15}$$

Denoting the polynomial obtained from $D(z_1, z_2)$ using the bilinear transformation as $Q(s_1, s_2)$ [we neglect the factor $(1 - s_1)^{m_1}(1 - s_2)^{n_1}$ that comes in the denominator], we can show that, for the case of QP filters [3.4]

$$D(z_1, z_2) \neq 0 \text{ in } \bar{U}^2$$

$$\Leftrightarrow \begin{cases} Q(s_1, s_2) \neq 0 & \text{for} \quad \text{Re}\{s_1\} \geqq 0, \text{Re}\{s_2\} \geqq 0 \\ s_2^{-n_1} Q(s_1, s_2)|_{s_2 = \infty} \neq 0 & \text{for} \quad \text{Re}\{s_1\} \geqq 0 \\ s_1^{-m_1} Q(s_1, s_2)|_{s_1 = \infty} \neq 0 & \text{for} \quad \text{Re}\{s_2\} \geqq 0 \\ s_1^{-m_1} s_2^{-n_1} Q(s_1, s_2)|_{s_1 = s_2 = \infty} \neq 0 & \end{cases} \tag{3.16}$$

and for the case of $\oplus +$ NSHP filters

$$D(z_1,z_2)\neq 0 \text{ in } \bar{U}^{20}$$

$$\Leftrightarrow \begin{cases} Q(s_1,s_2)\neq 0 & \text{for} \quad \mathrm{Re}\{s_1\}=0, \mathrm{Re}\{s_2\}\geqq 0 \\ s_2^{-n_1}Q(s_1,s_2)|_{s_2=\infty}\neq 0 & \text{for} \quad \mathrm{Re}\{s_1\}=0 \\ s_1^{-m_1}Q(s_1,s_2)|_{s_1=\infty}\neq 0 & \text{for} \quad \mathrm{Re}\{s_2\}\geqq 0 \\ s_1^{-m_1}s_2^{-n_1}Q(s_1,s_2)|_{s_1=s_2=\infty}\neq 0, & \end{cases} \tag{3.17a}$$

$$D(z_1,\infty)\neq 0 \quad \text{for} \quad \{|z_1|\geqq 1\}\Leftrightarrow Q(s_1,1)=(1-s_1)^{m'_1}Q_1(s_1), \tag{3.17b}$$

where

$$Q_1(s_1)\neq 0 \quad \text{for} \quad \mathrm{Re}\{s_1\}\geqq 0. \tag{3.18}$$

A polynomial $Q(s_1,s_2)$ having the property $Q(s_1,s_2)\neq 0$ for $\{\mathrm{Re}\{s_1\}\geqq 0, \mathrm{Re}\{s_2\}\geqq 0\}$ is known as a two-variable (2V) strictly-Hurwitz polynomial or 2V Hurwitz polynomial in the narrow sense [3.5] (the definition of 2V strictly-Hurwitz polynomial given here is slightly different from that given in the reference).

The bilinear transformation has found widespread acceptance because more than a decade ago, *Ansell* [3.5] showed that the 2V strictly-Hurwitz property of a polynomial can be tested in a finite number of steps. Therefore, we can use the bilinear transformation and *Ansell*'s test to check the stability of a 2D causal filter [3.6]. This should be compared with the original test for causal filter stability in the z_1, z_2 plane [3.7]. Here, one has to map the region $|z_2|\geqq 1$ onto the z_1 plane by the relation $D(z_1,z_2)=0$. The filter is stable only if the obtained values of $|z_1|$ [by solving $D(z_1,z_2)=0$] are less than one. This is a horrendous operation since we have to consider an infinite region of z_2 and solve for resulting values of z_1. There is another major reason for the interest in the bilinear transformation. *Ansell* [3.5] and others [3.8, 9] have demonstrated long ago that there is a definite relationship between 2V passive networks and 2V Hurwitz polynomials. This relationship has been exploited by the authors for the design of QP recursive filters and more recently for the design of HP filters. We shall come to this point later in this chapter.

3.2 Design of QP Recursive Filters

If the input to an LSI system is given by $x(m,n)=\exp[\mathrm{j}(\omega_1 m+\omega_2 n)]$, then from (3.2), we obtain the output as

$$\begin{aligned} y(m,n) &= \sum_{k,l}\sum h(k,l)\,\mathrm{e}^{\mathrm{j}\{(m-k)\omega_1+(n-l)\omega_2\}} \\ &= \mathrm{e}^{\mathrm{j}(m\omega_1+n\omega_2)}\sum_{k,l}\sum h(k,l)\,\mathrm{e}^{-\mathrm{j}(k\omega_1+l\omega_2)} \\ &= x(m,n)\,H(\mathrm{j}\omega_1,\mathrm{j}\omega_2). \end{aligned} \tag{3.19}$$

Where $H(j\omega_1, j\omega_2)$ is known as the *2D frequency response* of the LSI system and is the value of the system function evaluated on the unit bidisc $T^2 \equiv \{|z_1| = |z_2| = 1\}$. The frequency response is *continuous* and *periodic* in ω_1 and ω_2 with a period of 2π. Also, for real $h(m,n)$, $H(j\omega_1, j\omega_2)$ has certain symmetry properties with respect to the origin $\omega_1 = \omega_2 = 0$. They are that $|H(j\omega_1, j\omega_2)|$ is even and $\underline{/H(j\omega_1, j\omega_2)}$ is odd. Therefore, we can restrict ourselves to two finite quadrants of the ω_1 and ω_2 plane ($-\pi \leqq \omega_1 \leqq \pi$ and $0 \leqq \omega_2 \leqq \pi$ or $0 \leqq \omega_1 \leqq \pi$ and $-\pi \leqq \omega_2 \leqq \pi$) when considering the frequency response of an LSI system having *real* impulse response $h(m,n)$.

The normal objective of linear 2D digital filter design is to alter the frequency spectrum $X(j\omega_1, j\omega_2)$ of an input signal $x(m,n)$ by multiplying the spectrum by some frequency response function $H_D(j\omega_1, j\omega_2)$. The frequency domain digital filter design problem then consists of designing a filter with a frequency response $H(j\omega_1, j\omega_2)$ that approximates the desired frequency response function $H_D(j\omega_1, j\omega_2)$. Equivalently, the design problem may be carried out in the spatial domain wherein the filter is designed such that its impulse response $h(m,n)$ approximates the desired impulse response $h_D(m,n)$. In this section and the following one, we consider the frequency domain approximation of the given response $H_D(j\omega_1, j\omega_2)$ using causal QP filters[2] with transfer functions as given in (3.6). In terms of the transfer function, the design problem is then the selection of the coefficients $a(m,n)$ and $d(m,n)$ such that $H(j\omega_1, j\omega_2)$ approximates $H_D(j\omega_1, j\omega_2)$ in some sense. The "measure of closeness" of the approximation might be, for example, the least pth norm defined by

$$\|e\|_p \equiv \left\{ \int_{\omega_1} \int_{\omega_2} \{|H_D(j\omega_1, j\omega_2)| - |H(j\omega_1, j\omega_2)|\}^p d\omega_1 d\omega_2 \right\}^{1/p}; 1 \leqq p \leqq \infty . \tag{3.20}$$

The coefficients $a(m,n)$ and $d(m,n)$ are chosen such that $\|e\|_p$ is minimized. When the value of p in (3.20) tends to infinity, we obtain the Chebyshev (or minimax) norm of the error function.

There are many points worth noting here. First of all, we note that the frequency response is a continuous function of ω_1 and ω_2 and therefore we have to discretize the region $0 \leqq \omega_1 \leqq \pi$ and $-\pi \leqq \omega_2 \leqq \pi$ to be able to apply numerical techniques. The corresponding error function is then given by

$$\|e\|_p \equiv \left[\sum_I \sum_J \{|H_D\{j\omega_1(I), j\omega_2(J)\}| - |H\{j\omega_1(I), j\omega_2(J)\}|\}^p \right]^{1/p}, \tag{3.21}$$

where the indices I, J are over some finite subset of the ω_1, ω_2 plane. Since the error function is nonlinear in the design parameters, we have to resort to nonlinear optimization techniques to minimize it and such a minimization is common to many engineering problems. However, as we noted in the last section, *the coefficients $d(m,n)$ have to be constrained suitably to ensure stability*

2 We can obtain other types of QP filters, namely, $+-$, $-+$ and $--$ filters simply by making a transformation of varialbes $z_1 \Rightarrow z_1^{-1}$ and or $z_2 \Rightarrow z_2^{-1}$ in the transfer function of the causal filter.

and this complicates the recursive filter design problem. The testing of stability at each iteration of the nonlinear optimization process is too complicated and time consuming if not altogether impossible. (The reader is referred to Chap. 4 for a detailed discussion on stability of recursive filters.)

3.2.1 Direct Optimization Technique

In this subsection we briefly discuss a direct optimization technique for the design of QP recursive filters. For complete details, the reader is referred to reference [3.10].

The method is restricted to the design of 2D recursive filters consisting of first-order and second-order sections connected in cascade. That is, the transfer function is restricted to the form

$$H(z_1,z_2)=A_0\frac{\prod_{l=1}^{k}\left(\sum_{m=0}^{m_{2l}}\sum_{n=0}^{n_{2l}}a_l(m,n)z_1^{-m}z_2^{-n}\right)}{\prod_{l=1}^{k}\left(\sum_{m=0}^{m_{1l}}\sum_{n=0}^{n_{1l}}d_l(m,n)z_1^{-m}z_2^{-n}\right)}\,;\quad a_l(0,0)=d_l(0,0)\equiv 1\,, \tag{3.22}$$

where m_{1l}, m_{2l}, n_{1l} and n_{2l} can take the values of one and two.

Let us define $\Phi=[a_l(m,n),\, d_l(m,n),\, A_0]$ as the parameter vector and

$$J(\Phi)=\sum_I\sum_J[M\{\mathrm{j}\omega_1(I),\mathrm{j}\omega_2(J)\}-|H\{\mathrm{j}\omega_1(I),\mathrm{j}\omega_2(J)\}|]^p;\ p \text{ even} \tag{3.23}$$

as the performance index where $M\{\mathrm{j}\omega_1(I),\mathrm{j}\omega_2(J)\}$ are the required magnitude characteristics. Then the design proceeds as follows:

1) Assume initial values for the elements of the parameter vector Φ. The initial values for Φ should be such that the filter is stable (this can be easily achieved by assuming the denominator polynomials to be separable functions in z_1 and z_2).

2) Calculate the magnitude frequency response of the filter obtained.

3) Compute the performance index $J(\Phi)$.

4) Calculate the new step size $\Delta\Phi$ for the parameter values using an unconstrained optimization (refer to [3.10] for details and to [3.11] for some comments regarding the convergence of the algorithm) and update the parameter vector Φ to $\Phi+\Delta\Phi$.

5) Examine the stability of the filter obtained. If the filter is found to be unstable, successively reduce by half the incremental change $\Delta\Phi$ in the coefficients of the unstable sections until the filter becomes stable.

6) Repeat steps 2 through 5 until the approximation is satisfied or a certain number of iterations has been completed.

Remarks

The authors have selected the cascaded structure because it has the following advantages over the direct form:

i) The word length that guarantees the stability of the filter after the rounding of the coefficients is shorter. This is due to the fact that the sensitivity of the mapping of the unit circle of the z_2 plane into the z_1 plane under the denominator of the transfer function is lower.

ii) The error caused by quantization effect and finite coefficient size is lower.

iii) In 2D filters, the number of multiplications needed for each output using the cascade form is less than that needed when using the direct form.

iv) It is easier to check the stability of the cascade filter during the optimization process and make necessary changes in the coefficients of the unstable section to restore stability.

The method has the following drawbacks:

v) The stability has to be tested for each iteration of the algorithm and if an unstable section has been found, the step size has to be reduced and the stability has to be tested again. The process has to be repeated until all the sections are stable. This will have the effect of slowing down the optimization process and may lead to excessive computation time.

vi) It is not possible to represent a general higher order 2D polynomial as the product of first-order or second-order polynomials. Therefore the filters obtained will be only suboptimum.

Example 1: First-Order 2D Highpass Filter Design

In this example, we consider the design of a linear highpass filter with the following specifications

$$M\{j\omega_1(I), j\omega_2(J)\} = [\omega_1^2(I) + \omega_2^2(J)]^{1/2}/\pi$$

where[3]

$$\omega_1(I) = 0, \quad \frac{\pi}{14}, \ldots, \pi$$

$$\omega_2(J) = 0, \quad \frac{\pi}{14}, \ldots, \pi .$$

This filter is approximated by one first-order symmetric $[a(m,n) = a(n,m)$ and $d(m,n) = d(n,m)]$ section. The value of p is set to 2 in (3.23) and the transfer

3 Since the response is circularly symmetric, the sample points are restricted to the first quadrant of ω_1 and ω_2.

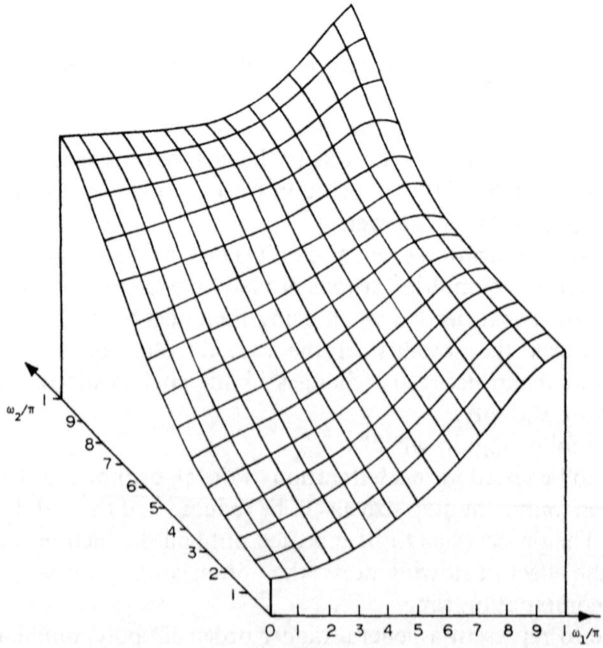

Fig. 3.4. Magnitude response of the 2D high-pass filter in example 1

function obtained by numerical optimization is given by

$$H(z_1, z_2) = 0.59944 \frac{[z_1 \ \ 1]\begin{bmatrix} 1 & -0.04668 \\ -0.04668 & -0.46761 \end{bmatrix}[z_2 \ \ 1]}{[z_1 \ \ 1]\begin{bmatrix} 1 & 0.42602 \\ 0.42602 & 0.10692 \end{bmatrix}[z_2 \ \ 1]}.$$

The amplitude response of the filter (for one quadrant) is shown in Fig. 3.4.

3.2.2 Design Using Transformed Variables

Here the design of a class of *stable* QP filters is considered. The stability is ensured in this method by the variable substitution procedure.

The use of transformation of variables or variable substitution procedure is well known to circuit theorists as a method for converting a constrained minimization technique (in which the parameter values are bounded) to an unconstrained minimization technique. For example, if $\{a_1, a_2, a_3, a_4\}$ are the variables of a constrained optimization procedure in which their values are to be constrained as given below

$$a_1>0; \quad a_2\geqq a_0=\text{constant}; \quad -1\leqq a_3\leqq 1 \quad \text{and} \quad 0<a_4<1$$

then we can use the following transformations

$$a_1=e^{b_1}; \quad a_2=a_0+b_2^2; \quad a_3=\sin(b_3) \quad \text{and} \quad a_4=\frac{e^{b_4}}{1+e^{b_4}}$$

and an unconstrained optimization to perform the minimization. The coefficients $\{b_1,b_2,b_3,b_4\}$ will be the varialbes of optimization in such a minimization and their values can be unbounded.

The method presented here is similar in spirit to the preceding principles. It can be used to design only a restricted class of causal filters. Letting $H(z_1,z_2)=A(z_1,z_2)/D(z_1,z_2)$ be the transfer function, we already noted in Sect. 3.1.4 that the stability constraint $D(z_1,z_2)\neq 0$ in $\bar{U}^2$ is equivalent to requiring that $Q(s_1,s_2)\neq 0$ for ($\text{Re}\{s_1\}\geqq 0$, $\text{Re}\{s_2\}\geqq 0$) and where $Q(s_1,s_2)$ is obtained from $D(z_1,z_2)$ using double bilinear transformation. Furthermore, it is known [3.5] that the above condition is equivalent to

$$Q(s_1,1)\neq 0 \quad \text{for} \quad \text{Re}\{s_1\}\geqq 0, \tag{3.24a}$$

$$Q(j\omega,s_2)\neq 0 \quad \text{for} \quad \text{Re}\{s_2\}\geqq 0 \quad \text{and for all real } \omega. \tag{3.24b}$$

Therefore, we can apply these two conditions to simple polynomials $Q(s_1,s_2)$ and obtain bounds on the values of their coefficients. Upon transformation of variables, we obtain the polynomials $Q(s_1,s_2)$ which have the strictly-Hurwitz property for all real values of the new variables. For example, it can be shown[4] [3.12] that the first-order (in each variable) polynomial

$$Q_l(s_1,s_2)=s_1s_2+a_{1l}^2s_1+a_{2l}^2s_2+a_{3l}^2; \quad a_{1l}\neq a_{2l}\neq a_{3l}\neq 0 \tag{3.25}$$

and the restricted second-order polynomial

$$Q_l(s_1,s_2)=\left(s_1s_2+a_{1l}^2s_1+a_{2l}^2s_2+\frac{a_{3l}^2+a_{1l}^2a_{4l}^2/4}{a_{2l}^2}+ja_{1l}^2a_{4l}\right)$$
$$\cdot\left(s_1s_2+a_{1l}^2s_1+a_{2l}^2s_2+\frac{a_{3l}^2+a_{1l}^2a_{4l}^2/4}{a_{2l}^2}-ja_{1l}^2a_{4l}\right);$$
$$\cdot a_{1l}\neq a_{2l}\neq a_{4l}\neq 0 \tag{3.26}$$

4 Some of the strictly-Hurwitz polynomials derived in [3.12] do not satisfy the additional constraints on $Q(s_1,s_2)$ as given in (3.16). Therefore, they will not lead to stable transfer functions $H(z_1,z_2)$.

satisfy the conditions given in (3.24) for all real values of $\{a_{il}\}$, that is $Q_l(s_1, s_2)$ remains strictly-Hurwitz for all real values of $\{a_{il}\}$. Therefore the transfer functions of the causal filters can be represented as

$$H(z_1, z_2) = A_0 \prod_{l=1}^{k} \frac{P_l(s_1, s_2)}{Q_l(s_1, s_2)}; \qquad s_i = \frac{z_i - 1}{z_i + 1} \qquad i = 1, 2, \tag{3.27}$$

where $Q_l(s_1, s_2)$ takes the form as given in (3.25, 26) and $P_l(s_1, s_2)$ is a polynomial in s_1, s_2 with degree (in each variable) less than or equal to that of $Q_l(s_1, s_2)$ for all l. Since $Q_l(s_1, s_2)$ retains its strictly-Hurwitz property for all real values of the variables $\{a_{il}\}$ and the coefficients of the numerator polynomials, we can use these variables and a standard unconstrained optimization to obtain a transfer function $H(z_1, z_2)$ having the required frequency response and the stability of $H(z_1, z_2)$ will be guaranteed at each iteration of the algorithm.

Examples of filters designed using the above approach can be found in [3.12].

Remarks

It should be pointed out that this method is somewhat superior to the direct optimization technique as the stability of the filter is always guaranteed. However, the filters obtained by this method are also suboptimum.

3.3 Design of Higher-Order Recursive Filters [3.14, 15]

In Sect. 3.2, we considered two methods for the design of QP recursive filters. Here we present a method for the design of QP filters that are guaranteed to be stable. The method can be used to obtain arbitrary order causal filter transfer functions. That is

$$H(z_1, z_2) = \frac{A(z_1, z_2)}{D(z_1, z_2)} = \frac{\sum_{m=0}^{m_2} \sum_{n=0}^{n_2} a(m, n) z_1^{-m} z_2^{-n}}{\sum_{m=0}^{m_1} \sum_{n=0}^{n_1} d(m, n) z_1^{-m} z_2^{-n}}, \tag{3.28}$$

where m_1 and n_1 are arbitrary integers and where $D(z_1, z_2) \neq 0$ in $\bar{U}^2$ due to stability requirements. The objective of the design procedure is to obtain a set of coefficients $a(m, n)$ and $d(m, n)$ such that $H(z_1, z_2)$ has the desired frequency response and the filter is stable.

The general 2D stability problem is solved here by transforming the coefficients $d(m, n)$ into functions of some other set of real coefficients $\{y_{kl}\}$, $l > k$ where k and l are integers in some finite range $2 \leqq k, l \leqq n$. Sufficiency conditions are established such that for all real values of $\{y_{kl}\}$ the corresponding transfer functions obtained are stable. That is

$$\{d(m,n)\}=\mathrm{Fn}[y_{kl}]; \quad \text{Fn denotes Function of} \tag{3.29}$$

and the corresponding denominator polynomials obtained using the above transformation have the property

$$D(z_1,z_2)\neq 0 \quad \text{in} \quad \bar{U}^2 \quad \text{for all real } \{y_{kl}\}. \tag{3.30}$$

The method presented here is based on:

i) the property of the polynomial $Q(s_1,s_2)$ obtained from $D(z_1,z_2)$ using the bilinear transformation and

ii) some well-known results in 2V passive network theory.

We considered i) in detail in Sect. 3.1.4 and as noted there

$$D(z_1,z_2)\neq 0 \quad \text{in} \quad \bar{U}^2 \Rightarrow Q(s_1,s_2)\neq 0 \quad \text{for} \quad (\mathrm{Re}\{s_1\}\geqq 0, \mathrm{Re}\{s_2\}\geqq 0). \tag{3.31}$$

Therefore, if we find a way to obtain the polynomial $Q(s_1,s_2)$ of arbitrary degree which remains strictly-Hurwitz for all real values of the above-mentioned coefficients $\{y_{kl}\}$, then we can obtain stable denominator polynomials $D(z_1,z_2)$ from $Q(s_1,s_2)$ using the bilinear transformation.

Some properties of 2V passive networks are used in this section to generate strictly-Hurwitz polynomials of arbitrary degree. A 2V passive network is a proper interconnection of reactive elements (inductors and capacitors) in the complex variables s_1 and s_2 and frequency independent elements such as resistors, gyrators and transformers. A 2V reactive or lossless network is one which is void of resistors. The properties of such network are well known. Here, we invoke the particular property that *the sum of the numerator and the denominator polynomial of the driving point immittance function of a 2V lossless network is 2V strictly-Hurwitz* [3.5, 9, 13]. That is, if we have a 2V lossless network as shown in Fig. 3.5 and define the input admittance function at port 1 as

$$Y_{in}(s_1,s_2)=\frac{M(s_1,s_2)}{N(s_1,s_2)}, \tag{3.32}$$

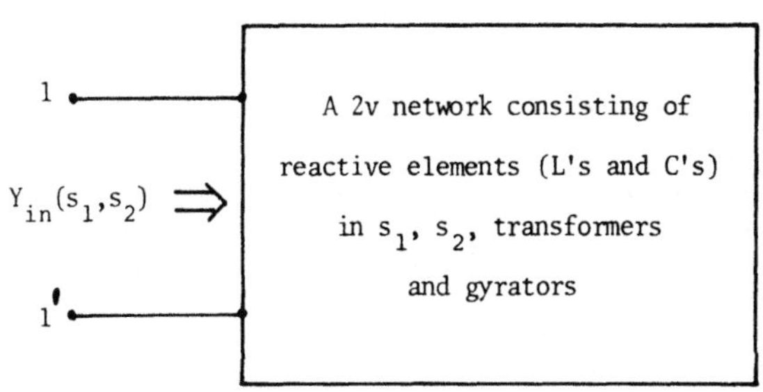

Fig. 3.5. A 2V lossless network

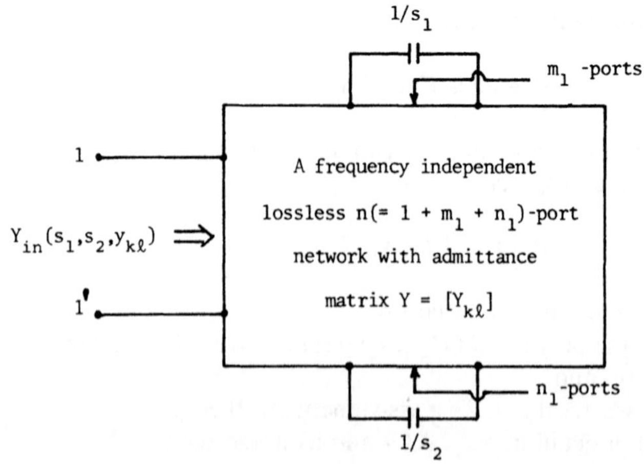

Fig. 3.6. A 2V lossless with $m_1(n_1)$ unit valued $s_1(s_2)$ capacitors

where $M(s_1, s_2)$ and $N(s_1, s_2)$ are mutually prime polynomials, then

$$Q(s_1, s_2) = M(s_1, s_2) + N(s_1, s_2) \tag{3.33}$$

is strictly-Hurwitz[5]. More importantly, $Q(s_1, s_2)$ retains its strictly-Hurwitz nature for all possible values of the network elements.

From the preceding discussion, we can conclude that if we have a 2V lossless network in which the elements can take on any real values, then we can use the polynomial $Q(s_1, s_2)$ obtained from such a network for the purpose of generating stable QP transfer functions. We have given one such network in Fig. 3.6. That network contains m_1 unit capacitors in the complex variable s_1 and n_1 unit capacitors in s_2. The inside block which is a $n(=m_1+n_1+1)$-port network is lossless and frequency independent. We can describe this network in terms of its admittance matrix Y[6]. Because the network is, by definition, lossless and frequency independent, it immediately follows that all the elements of Y are real and Y is skew symmetric. We can therefore write Y as

$$\mathrm{Y} = \left|\begin{array}{c|ccccc} 0 & y_{12} & y_{13} & \cdot & \cdot & y_{1n} \\ \hline -y_{12} & 0 & y_{23} & \cdot & \cdot & y_{2n} \\ -y_{13} & -y_{23} & 0 & \cdot & \cdot & \cdot \\ \cdot & \cdot & \cdot & \cdot & \cdot & \cdot \\ \cdot & \cdot & \cdot & \cdot & \cdot & \cdot \\ -y_{1n} & -y_{2n} & \cdot & & \cdot & 0 \end{array}\right| = \left|\begin{array}{c|c} 0 & Y_{12} \\ \hline -Y_{12}^t & Y_{22} \end{array}\right|. \tag{3.34}$$

5 We should point out here that $Q(s_1, s_2)$ obtained using this method might sometimes have zeros at finite number of points for which $\mathrm{Re}\{s_1\} = \mathrm{Re}\{s_2\} = 0$. This condition happens when $Y_{in}(s_1, s_2)$ has nonessential singularities of the second kind on $\mathrm{Re}\{s_1\} = \mathrm{Re}\{s_2\} = 0$. We will come to this point later in this section.

6 We assume here that an admittance matrix representation exists. However, this in no way affects the validity of the method, as we can always proceed using a scattering matrix representation.

Table 3.1. 2V strictly-Hurwitz polynomials

m_1	n_1	n	$Q(s_1, s_2, y_{kl})$
1	1	3	$s_1 s_2 + y_{13}^2 s_1 + y_{12}^2 s_2 + y_{23}^2$
2	1	4	$s_1^2 s_2 + y_{14}^2 s_1^2 + (y_{12}^2 + y_{13}^2) s_1 s_2 + (y_{24}^2 + y_{34}^2) s_1 + y_{23}^2 s_2 + (y_{12}y_{34} - y_{13}y_{24} + y_{14}y_{23})^2$
2	2	5	$s_1^2 s_2^2 + (y_{12}^2 + y_{13}^2) s_1 s_2^2 + (y_{14}^2 + y_{15}^2) s_1^2 s_2 + y_{45}^2 s_1^2 + y_{23}^2 s_2^2 + (y_{24}^2 + y_{25}^2 + y_{34}^2 + y_{35}^2) s_1 s_2 + [(y_{12}y_{45} - y_{14}y_{25} + y_{15}y_{24})^2 + (y_{13}y_{45} - y_{14}y_{35} + y_{15}y_{34})^2] s_1 + [(y_{12}y_{34} - y_{13}y_{24} + y_{14}y_{23})^2 + (y_{12}y_{35} - y_{13}y_{25} + y_{15}y_{23})^2] s_2 + (y_{23}y_{45} - y_{24}y_{35} + y_{25}y_{34})^2$

Defining $\hat{Y}_{22}(s_1, s_2, y_{kl})$ as

$$\hat{Y}_{22}(s_1, s_2, y_{kl}) = Y_{22} + \text{dia}[\overbrace{s_1 \ \ldots \ s_1}^{m_1}\overbrace{s_2 \ \ldots s_2}^{n_1}] \tag{3.35a}$$

and $\hat{Y}(s_1, s_2, y_{kl})$ as

$$\hat{Y}(s_1, s_2, y_{kl}) = Y + \text{dia}[\overbrace{s_1 \ \ldots \ s_1}^{m_1}\overbrace{s_2 \ \ldots s_2}^{n_1}] \tag{3.35b}$$

we can show that the input admittance at port 1 of this network is given by

$$Y_{in}(s_1, s_2, y_{kl}) = \frac{M(s_1, s_2, y_{kl})}{N(s_1, s_1, y_{kl})} \equiv \frac{Y_{12}[\text{Adj}\{\hat{Y}_{22}(s_1, s_2, y_{kl})\}]\, Y_{12}^t}{\text{Det}[\hat{Y}_{22}(s_1, s_2, y_{kl})]} \tag{3.36}$$

and the polynomial

$$\begin{aligned} Q(s_1, s_2, y_{kl}) &= M(s_1, s_2, y_{kl}) + N(s_1, s_2, y_{kl}) \\ &= \text{Det}[Y(s_1, s_2, y_{kl})] \end{aligned} \tag{3.37}$$

which is a function of $\{y_{kl}\}$, remains strictly-Hurwitz for all real values of $\{y_{kl}\}$. In Table 3.1, the polynomial $Q(s_1, s_2, y_{kl})$ for the cases m_1, $n_1 = 1, 2$ is given.

We now have a polynomial $Q(s_1, s_2, y_{kl})$ which is strictly-Hurwitz. The essential feature of the method is then to rewrite the transfer function of the QP recursive filters as

$$H(z_1, z_2) = \left. \frac{A(z_1, z_2)}{Q(s_1, s_2, y_{kl})\,(z_1+1)^{m_1}\,(z_2+1)^{n_1}} \right|_{s_i = \frac{z_i - 1}{z_i + 1},\, i = 1, 2} \tag{3.38}$$

and use the coefficients $a(m, n)$ and $\{y_{kl}\}$ as the variables of an optimization procedure. Since $Q(s_1, s_2, y_{kl})$ remains strictly-Hurwitz for all real values of $\{y_{kl}\}$

and because of the property of the bilinear transformation, the stability of $H(z_1, z_2)$ *is always guaranteed.* Also because the element values can range from $-\infty$ to ∞, we can use a standard unconstrained optimization to minimize a suitable error function.

Remarks

1) As noted in Sect. 3.1.4, the strictly-Hurwitz nature of a polynomial $Q(s_1, s_2, y_{kl})$ is *not* sufficient to ensure the min–min phase property of the polynomial $D(z_1, z_2)$ obtained using the bilinear transformation. Specifically, if

$$Q(s_1, s_2, y_{kl}) = \sum_{m=0}^{m_1} \sum_{n=0}^{n_1} q(m, n) s_1^m s_2^n . \tag{3.39}$$

Then, we also require[7]

$$\begin{aligned} Q_1(s_1) &\equiv s_2^{-n_1} Q(s_1, s_2, y_{kl})|_{s_2=\infty} \\ &= \sum_{m=0}^{m_1} q(m, n_1) s_1^m \neq 0 \quad \text{in} \quad \mathrm{Re}\{s_1\} \geqq 0 \end{aligned} \tag{3.40}$$

and

$$\begin{aligned} Q_2(s_2) &\equiv s_1^{-m_1} Q(s_1, s_2, y_{kl})_{s_1=\infty} \\ &= \sum_{n=0}^{n_1} q(m_1, n) s_2^n \neq 0 \quad \text{in} \quad \mathrm{Re}\{s_2\} \geqq 0 . \end{aligned} \tag{3.41}$$

We now prove that the polynomials $Q(s_1, s_2, y_{kl})$ obtained from the 2V lossless network of Fig. 3.6 do indeed have the above properties.

Proof. Considering the network in Fig. 3.6, we can note that the polynomials $Q_1(s_1)$ and $Q_2(s_2)$, as defined above, correspond exactly to the sum of the numerator and the denominator of the 1V driving point admittance function (at port 1) obtained by shorting, respectively, the s_2 and s_1 capacitors. Therefore these polynomials will automatically have the properties given above.

2) This method can be used to design *all* QP filter transfer functions $H(z_1, z_2)$ whose denominator polynomial, upon the bilinear transformation, can be represented as

$$Q(s_1, s_2) = [1 \; s_1 \ldots s_1^{m_1}] \begin{bmatrix} q(0,0) & q(0,1) & . & . & . & . \\ q(1,0) & . & . & . & . & . \\ . & . & . & . & . & . \\ . & . & . & . & . & . \\ . & . & . & . & . & . \\ . & . & . & . & . & 1 \end{bmatrix} \begin{bmatrix} 1 \\ s_2 \\ . \\ . \\ . \\ s_2^{n_1} \end{bmatrix} \tag{3.42}$$

7 Note that $Q(s_1, s_2, y_{kl})$ obtained from the network of Fig. 3.6 satisfies the condition $s_1^{-m_1} s_2^{-n_1} Q(s_1, s_2, y_{kl})|_{s_1 \; s_2=\infty} \neq 0$ for all real values of $\{y_{kl}\}$.

and where all elements $q(m,n)>0$. This simply follows because of the following reasons:

i) There corresponds a 2V strictly-Hurwitz polynomial $Q(s_1,s_2)$ for each min–min phase polynomial.

ii) The ratio of odd to even part of any 2V strictly-Hurwitz polynomial is a 2V rectance function [3.5].

iii) Every reactance function is realizable as the immittance function of a 2V reactive network [3.16], where the network contains the minimum number of reactive elements in s_1 and s_2. [This minimum number is given by the degree of $Q(s_1,s_2)$ in s_1 and s_2.] If we pull out the reactive elements from such a minimal realization we will in fact obtain a 2V reactive network as shown in Fig. 3.6, proving our statement.

3) As mentioned before, there may be a set of values of $\{y_{kl}\}$ which will lead to a polynomial $Q(s_1,s_2,y_{kl})$ having zeros at $\mathrm{Re}\{s_1\}=\mathrm{Re}\{s_2\}=0$. As a specific example, let us consider the polynomial $Q(s_1,s_2,y_{kl})$ which is of degree 2 in $s_1(m_1=2)$ and 1 in $s_2(n_1=1)$ obtained when $\{y_{12},y_{13},y_{14},y_{23},y_{24},y_{34}\}\equiv\{1, 1, 1, 1, -1, 1\}$.

We can show that

$$\begin{aligned}Q(s_1,s_2)&=[1\ s_1\ s_1^2]\begin{bmatrix}9&1\\2&2\\1&1\end{bmatrix}\begin{bmatrix}1\\ \\s_2\end{bmatrix}\\&=(9+2s_1s_2+s_1^2)+(s_2+2s_1+s_1^2s_2)\\&\equiv M(s_1,s_2)+N(s_1,s_2)\end{aligned}\tag{3.43}$$

and

$$\begin{aligned}Q(s_1,s_2)&\equiv 0 \quad \text{for} \quad \{s_1=s_2=\pm j\sqrt{3}\}\\&\not\equiv 0 \text{ elsewhere in } (\mathrm{Re}\{s_1\}\geqq 0, \mathrm{Re}\{s_2\}\geqq 0).\end{aligned}\tag{3.44}$$

This polynomial $Q(s_1,s_2)$ if used in (3.38) will lead to an unstable $H(z_1,z_2)$. The situation *would not occur during optimization* if the initial values of $\{y_{kl}\}$ are such that the polynomial $Q(s_1,s_2)$ is void of such zeros. We can also avoid this problem completely by replacing each capacitor in the network of Fig. 3.6 by a parallel combination of capacitors of unit value and resistors of value $\alpha_{1j}^2(j=1$ to m_1 for the s_1 capacitors) or $\alpha_{2j}^2(j=1$ to n_1 for the s_2 capacitors) and use the new $Q(s_1,s_2,y_{kl},\alpha_{ij})$ in (3.38). However, this results in an increase in the number of variables of optimization (m_1+n_1 in all) and, more importantly, may restrict the types of frequency responses that can be obtained. The latter becomes obvious if we note that we are in effect making frequency tranformations as given by

$$s_i\Rightarrow s_i+\alpha_i^2;\qquad i=1,2 \quad\text{and}\quad \alpha_i^2\equiv\min[\alpha_{ij}^2, j=1 \text{ to } m_1 \text{ or } n_1].$$

Because of these drawbacks and from our experience, we favor the original approach.

4) Since it is very difficult to obtain $Q(s_1, s_2, y_{kl})$ in a closed form, that is as functions of the elements $\{y_{kl}\}$, it becomes necessary to employ numerical techniques. Also, for $m_1, n_1 \gg 1$, the total number of variables of optimization $\{y_{kl}\}$ greatly exceeds that of $d(m,n)$. To bring down this total number, we can constrain some of the coefficients of $\{y_{kl}\}$ to zero. A suitable reduced matrix Y' is given by

$$Y' = \left[\begin{array}{cccccc|ccccc}
0 & y'_{12} & 0 & \cdot & \cdot & 0 & y'_{1k} & 0 & \cdot & \cdot & 0 \\
-y'_{12} & 0 & y'_{23} & \cdot & \cdot & \cdot & y'_{2k} & \cdot & \cdot & \cdot & y'_{2n} \\
0 & -y'_{23} & 0 & \cdot & \cdot & \cdot & \cdot & \cdot & \cdot & \cdot & \cdot \\
\cdot & \cdot & \cdot & \cdot & \cdot & 0 & \cdot & \cdot & \cdot & \cdot & \cdot \\
\cdot & \cdot & \cdot & \cdot & \cdot & y'_{ij} & \cdot & \cdot & \cdot & \cdot & \cdot \\
0 & \cdot & \cdot & 0 & -y'_{ij} & 0 & y'_{jk} & \cdot & \cdot & \cdot & y'_{jn} \\
\hline
-y'_{1k} & -y'_{2k} & \cdot & \cdot & \cdot & -y'_{jk} & 0 & y'_{kl} & 0 & \cdot & 0 \\
0 & \cdot & \cdot & \cdot & \cdot & \cdot & -y'_{kl} & 0 & \cdot & \cdot & \cdot \\
\cdot & \cdot & \cdot & \cdot & \cdot & \cdot & 0 & \cdot & \cdot & \cdot & 0 \\
\cdot & \cdot & \cdot & \cdot & \cdot & \cdot & \cdot & \cdot & \cdot & \cdot & y'_{mn} \\
0 & -y'_{2n} & \cdot & \cdot & \cdot & -y'_{jn} & 0 & \cdot & 0 & -y'_{mn} & 0
\end{array}\right] \qquad \begin{array}{l} i=m_1 \\ j=1+m_1 \\ k=2+m_1 \\ m=m_1+n_1 \\ n=1+m_1+n_1 \end{array} \tag{3.45}$$

This Y' matrix is used in (3.35–37) to obtain the 2V strictly-Hurwitz polynomial. The reader is referred to [3.14, 15] for details.

Example 2: Circularly Symmetric Lowpass Filter Design

In this example we present results on the design of a QP filter approximating a circularly symmetric lowpass magnitude characteristic given by

$$M[j\omega_1(I), j\omega_2(J)] = \begin{cases} 1 & \text{for} \quad \sqrt{\omega_1^2(I)+\omega_2^2(J)} \leq 0.2\pi \\ 0 & \text{otherwise}. \end{cases} \tag{3.46}$$

The transfer function is assumed to be of all pole type (in s_1, s_2) and the order in each variable is set to five ($m_1 = n_1 = 5$). Thus we have

$$H(z_1, z_2) = \frac{A_N}{Q\left(\frac{z_1-1}{z_1+1}, \frac{z_2-1}{z_2+1}, y'_{kl}\right)} = \frac{B_N(z_1+1)^5 (z_2+1)^5}{\sum\limits_{m=0}^{5} \sum\limits_{n=0}^{5} d(m,n)\, z_1^{5-m} z_2^{5-n}}, \tag{3.47}$$

Table 3.2. Coefficients of the fifth-order lowpass filter in example 2

$B_N = 0.28627$

$[d(m,n)]=$	0	1	2→n	3	4	5	
	78.428	140.20	101.47	−36.029	6.0739	−0.41341	0
	−128.43	136.76	−33.313	−7.4214	1.4820	0.79070	1
	83.011	−22.705	−7.2275	−33.487	28.707	−6.3054	2 ↓ m
	−25.519	−13.249	−29.302	61.551	−28.734	4.2941	3
	3.4667	3.4048	23.838	−25.871	7.8851	−0.79298	4
	−0.16445	0.32176	−4.8317	3.3632	−0.64500	0.065181	5

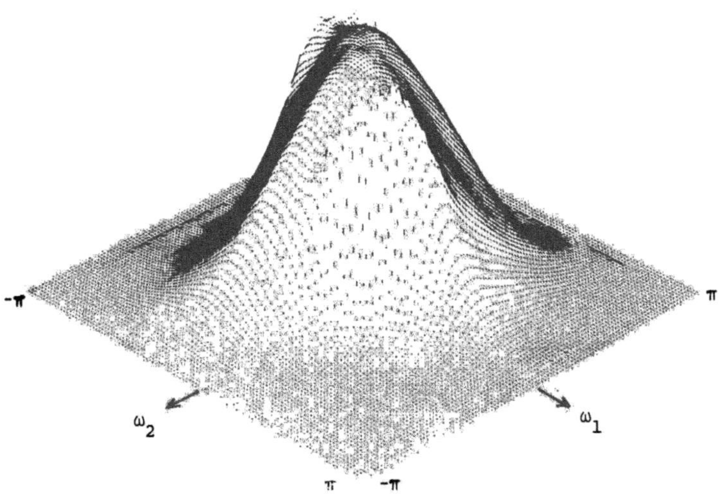

Fig. 3.7. Magnitude response of the 2D lowpass filter in example 2

where $Q(s_1, s_2, y'_{kl})$ is as defined in (3.37). The frequency sample points are selected as

$$\omega_1(I) = \begin{cases} 0.0, 0.2\pi(0.01\pi)^{8} & I = 0, 1, \ldots, 20 \\ 0.3\pi, 0.6\pi(0.1\pi) & I = 21, \ldots, 24 \end{cases}$$
$$\omega_2(24 \pm I) = \pm\omega_1(I) \qquad I = 0, 1, \ldots, 24. \tag{3.48}$$

The coefficients A_N and $\{y'_{kl}\}$ of the reduced Y' matrix in (3.45) are used as the variables of an unconstrained optimization minimizing the performance index $J(\Phi)$ given in (3.23) and where $p=2$. The algorithm that has been employed uses a quasi-Newton technique [3.17]. The final values of the coefficients B_N, $d(m,n)$

8 The notation $x = x_1, x_2(y)$ is used to denote that x takes the values of $x_1, x_1 + y, \ldots, x_2$.

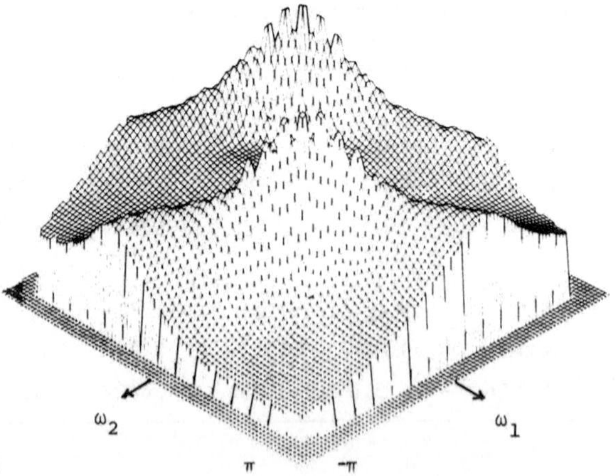

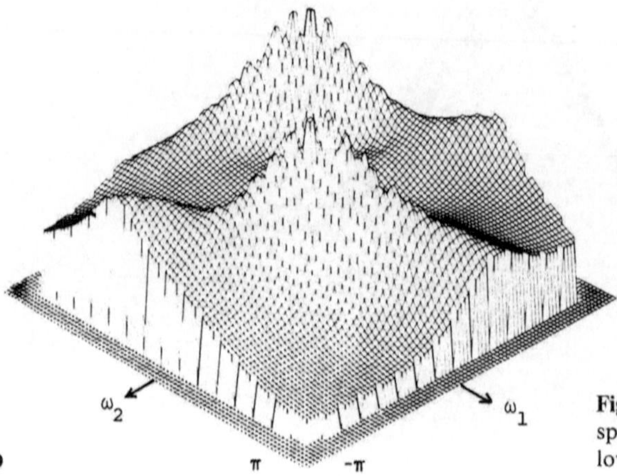

Fig. 3.8a, b. Group delay responses τ_1 and τ_2 of the 2D lowpass filter

are given in Table 3.2. A 3D plot of the magnitude response $|H(j\omega_1, j\omega_2)|$ is shown in Fig. 3.7. The group delay responses $\tau_1(j\omega_1, j\omega_2)$ and $\tau_2(j\omega_1, j\omega_2)$ of the filter and

$$\tau_i(j\omega_1, j\omega_2) = \frac{\partial}{\partial \omega_i}[\underline{/H(j\omega_1, j\omega_2)}] \quad i=1,2 \tag{3.49}$$

are shown in Fig. 3.8a, b.

Example 3: Allpass Filter Design

In the previous example we considered the design of a QP filter approximating a desired magnitude response. For this filter we can note that the group delays exhibit a large peak-to-peak variations near the band edges. In this example we try to equalize these large group delay variations and obtain overall flat group delay responses by cascading a recursive 2D allpass filter with the first filter[9]. We again use a transfer function of order 5 in each variable, given by

$$H(z_1,z_2)=\frac{Q\left(\frac{1-z_1}{z_1+1},\frac{1-z_2}{z_2+1},y'_{kl}\right)}{Q\left(\frac{z_1-1}{z_1+1},\frac{z_2-1}{z_2+1},y'_{kl}\right)}\equiv\frac{\sum_{m=0}^{5}\sum_{n=0}^{5}d(m,n)z_1^m z_2^m}{\sum_{m=0}^{5}\sum_{n=0}^{5}d(m,n)z_1^{5-m}z_2^{5-n}}. \tag{3.50}$$

The frequency sample points are restricted to those inside of the lowpass filter. Thus we have

$$\omega_1(I)=0, 0.2\pi(0.01\pi) \qquad I=0,1,\dots,20$$

$$\omega_2(20\pm I)=\pm\omega_1(I) \qquad I=0,1,\dots,20$$

and

$$\sqrt{\omega_1^2(I)+\omega_2^2(20\pm I)}\leqq 0.2\pi. \tag{3.51}$$

The coefficients $d(m,n)$ of the transfer function $H(z_1,z_2)$ are obtained by minimizing the objective function given by

$$\begin{aligned}J_{\mathrm{AP}}(\Phi)=\sum_I\sum_J[\{&\tau_{1\,\mathrm{spec}}(\mathrm{j}\omega_1(I),\mathrm{j}\omega_2(J))-\tau_{1\,\mathrm{LP}}(\mathrm{j}\omega_1(I),\mathrm{j}\omega_2(J))\\&-\tau_{1\,\mathrm{AP}}(\mathrm{j}\omega_1(I),\mathrm{j}\omega_2(J))-\tau_{10}\}^2\\&+\{\tau_{2\,\mathrm{spec}}(\mathrm{j}\omega_1(I),\mathrm{j}\omega_2(J))-\tau_{2\,\mathrm{LP}}(\mathrm{j}\omega_1(I),\mathrm{j}\omega_2(J))\\&-\tau_{2\,\mathrm{AP}}(\mathrm{j}\omega_1(I),\mathrm{j}\omega_2(J))-\tau_{20}\}^2],\end{aligned} \tag{3.52}$$

where $\tau_{1\,\mathrm{spec}}$ and $\tau_{2\,\mathrm{spec}}$ are the specified constant group delays in the direction of ω_1 and ω_2, respectively, $\tau_{1\,\mathrm{LP}}$, $\tau_{2\,\mathrm{LP}}$ are the group delays of the lowpass filter, $\tau_{1\,\mathrm{AP}}$, $\tau_{2\,\mathrm{AP}}$ are the group delays of the allpass filter and τ_{10}, τ_{20} are constants used as variables of optimization.

The same quasi-Newton technique is used to minimize this objective function using $\{y'_{kl}\}$, τ_{10} and τ_{20} as the variables of optimization. The resulting

---

9 We could have used the first filter itself to approximate simultaneously the magnitude and the group delay characteristics. This is simply achieved by adding suitable error functions for the group delays to the performance index $J(\Phi)$ in (3.23). In fact, *Aly* and *Fahmy* [3.18] have demonstrated recently that the polynomials $Q(s_1,s_2,y_{kl})$ given in Table 3.1 can be used effectively to design such filters.

Table 3.3. Coefficients of the fifht-order allpass filter in example 3

$[d(m, n)] =$

$m \downarrow$ / $n \rightarrow$	0	1	2	3	4	5
0	37.958	−66.016	59.643	−33.841	11.183	−1.6736
1	−62.890	70.765	− 55.457	37.714	−15.753	2.9720
2	51.780	−47.517	67.362	−66.084	30.783	−5.9439
3	−26.644	30.318	−636.17	57.765	−24.323	4.4651
4	7.9760	−11.965	27.010	−21.361	8.8920	−1.8426
5	− 1.0483	1.9329	− 4.3356	3.1629	− 1.6162	0.38218

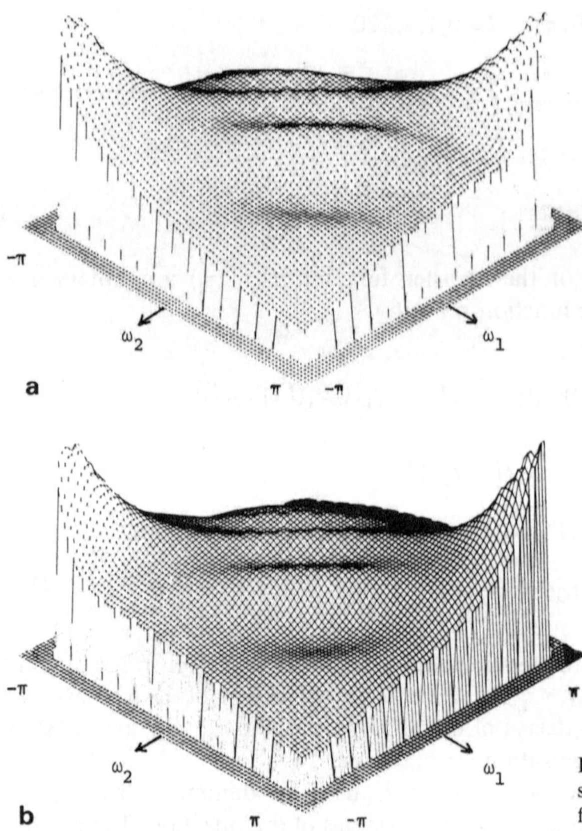

Fig. 3.9a, b. Group delay responses τ_1 and τ_2 of the 2D allpass filter in example 3

matrix $[d(m, n)]$ is given in Table 3.3. The group delays of the allpass filter and the overall system are shown in Figs. 3.9a, b and 10a, b, respectively. A comparison of Figs. 3.10a, b with 8a, b reveals a significant improvement in the group delay characteristics of the overall system.

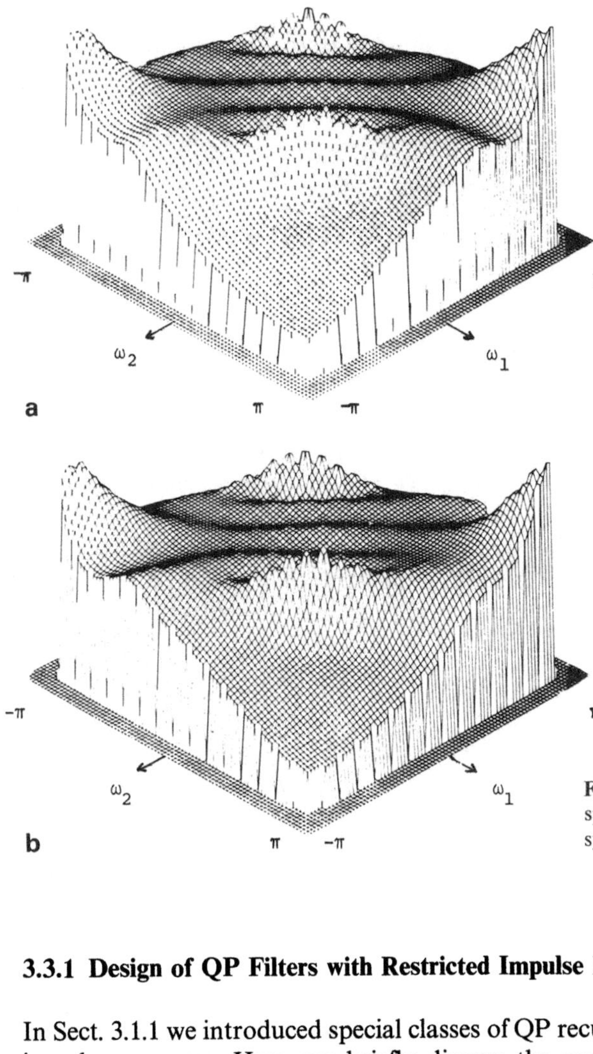

Fig. 3.10a, b. Group delay responses τ_1 and τ_2 of the cascaded system

3.3.1 Design of QP Filters with Restricted Impulse Response

In Sect. 3.1.1 we introduced special classes of QP recursive filters with restricted impulse response. Here, we briefly discuss the application of the preceding method of designing high-order QP filters to the design of such special classes of filters. Specifically, we consider the design of filters with an input-output relationship as given in (3.5) and where the region of support of the coefficients is $\mathscr{R}f^3_{++}$. The system function for this class of filters is given by

$$H(z_1,z_2)=\frac{A(z_1,z_2)}{D(z_1,z_2)}=\frac{\sum\sum_{\mathscr{R}f^3_{++}} a(m,n)\,z_1^{-m}z_2^{-n}}{\sum\sum_{\mathscr{R}f^3_{++}} d(m,n)\,z_1^{-m}z_2^{-n}}$$

$$\equiv h(0,0)+\sum_{m=1}^{\infty}\sum_{n=1}^{\infty} h(m,n)\,z_1^{-m}z_2^{-n}. \tag{3.53}$$

That is, the impulse response of this class of filters is zero on the open axes $\{m=0, n>0\}$ and $\{m>0, n=0\}$ and in addition to the stability requirement $D(z_1, z_2) \neq 0$ in $\bar{U}^2$, the denominator polynomial should have the property

$$D(z_1, \infty) \equiv D(\infty, z_2) \equiv d(0,0). \tag{3.54}$$

Now applying the bilinear transformation to $D(z_1, z_2)$ we can show that the above condition in terms of the polynomial $Q(s_1, s_2)$ is given by

$$Q(s_1, 1) \equiv k_1(1+s_1)^{m_1} \quad k_1>0, \tag{3.55a}$$

$$Q(1, s_2) \equiv k_2(1+s_2)^{n_1} \quad k_2>0. \tag{3.55b}$$

Therefore, $Q(s_1, s_2)$ in addition being 2V strictly-Hurwitz, should have $m_1(n_1)$ zeros at $s_1=-1(s_2=-1)$ when the other variable $s_2(s_1)$ is set to 1. If we find a method to generate such polynomials, we can use these polynomials to obtain stable transfer functions.

Now, consider the poynomial $Q(s_1, s_2, y_{kl})$ in (3.37) obtained from the 2V lossless network. We have already noted that $Q(s_1, s_2, y_{kl})$ remains strictly-Hurwitz for all real $\{y_{kl}\}$. Suppose we set $s_2=1$ in this polynomial and define the resulting polynomial as

$$Q_1(s_1, y_{kl}) \equiv Q(s_1, s_2=1, y_{kl}) = \sum_{m=0}^{m_1} q_1(m, y_{kl}) s_1^m. \tag{3.56}$$

We can note that setting $s_2=1$ corresponds to replacing all s_2 plane capacitors in the 2V network of Fig. 3.6 by unit valued resistors and $Q_1(s_1, y_{kl})$ corresponds to the sum of the numerator and the denominator of the admittance function of the resulting 1V passive network. Therefore it follows that

$$Q_1(s_1, y_{kl}) \neq 0 \quad \text{in} \quad \mathrm{Re}\{s_1\} \geq 0 \text{ and for all real } \{y_{kl}\}. \tag{3.57}$$

That is $Q_1(s_1, y_{kl})$ will have all its zeros (m_1 in all) in the open left half-plane of s_1. Therefore by constraining the coefficients $\{y_{kl}\}$ suitably, we can force $Q_1(s_1, y_{kl})$ to have m_1 zeros at $s_1=-1$. This is exactly the condition we have in (3.55a). A similar statement applies to $Q_2(s_2, y_{kl}) \equiv Q(s_1=1, s_2, y_{kl})$.

To summarize, we rewrite the transfer function of the QP filter as

$$H(z_1, z_2) = \left. \frac{A(z_1, z_2)}{Q(s_1, s_2, y_{kl})(z_1+1)^{m_1}(z_2+1)^{n_1}} \right|_{s_i = \frac{z_i-1}{z_i+1},\ i=1,2}, \tag{3.58}$$

where $Q(s_1, s_2, y_{kl})$ is as given in (3.37). The variables $a(m,n)$ and $\{y_{kl}\}$ are used as the variables of an optimization minimizing a suitable error function between the desired response and the actual response. The variables are constrained suitably such that (3.55a, b) are satisfied. [This can be done by another error function connecting the actual location of the zeros of $Q(s_1, 1, y_{kl})$ and $Q(1, s_2, y_{kl})$ with the desired zero location.]

3.4 Design of Half-Plane Recursive Filters

In the last two sections, we considered the design of QP filters. In this section, we consider the design of HP recursive filters and some examples of this class of filter are presented.

Referring back to Figs. 3.1b and 3b, it is observed that $++$ QP filters use *previously computed outputs* that lie *below and to the left* only, when calculating the present output whereas $\oplus +$ *NSHP filters use outputs that lie to the right also.* In terms of the present output $y(m, n)$, this means that the variable k used in the difference equation (3.8) of $\oplus +$ *NSHP filter can be allowed to take values greater than m* as long as the other variable l is less than n. That is, the $\oplus +$ filter is noncausal in the direction of m and causal in the direction of n (hence the name semicausal filter) whereas the $++$ QP filter is causal in both directions. This implies that $++$ QP filters form a subset of the more general $\oplus +$ NSHP filters where the latter are now shown to offer more flexibility in terms of the type of input-output response that can be obtained.

Consider a 2D autocorrelation or nonnegative definite function $r(m, n)$. Let $S(z_1, z_2)$ be its polynomial spectral function. Note

$$S(j\omega_1, j\omega_2) = S(z_1 = e^{j\omega_1}, z_2 = e^{j\omega_2}). \tag{3.59}$$

Then, as shown in [3.1], we can spectrally factor [10] $S(z_1, z_2)$ as

$$S(z_1, z_2) = D_{++}(z_1, z_2)\, D_{+-}(z_1, z_2)\, D_{--}(z_1, z_2)\, D_{-+}(z_1, z_2) \tag{3.60}$$

or equivalently $r(m, n)$ as

$$r(m, n) = d_{++}(m, n) * d_{+-}(m, n) * d_{--}(m, n) * d_{-+}(m, n). \tag{3.61}$$

Here, $D_{++}(z_1, z_2)$ is a min–min phase polynomial, $D_{+-}(z_1, z_2)$ is a min–max polynomial and so on and $d_{++}(m, n)$, etc., are infinite-order QP factors even if $r(m, n)$ is of finite order. Also, we have

$$D_{--}(z_1, z_2) = D_{++}(z_1^{-1}, z_2^{-1}), \tag{3.62a}$$

$$D_{-+}(z_1, z_2) = D_{+-}(z_1^{-1}, z_2^{-1}), \tag{3.62b}$$

and therefore

$$S(j\omega_1, j\omega_2) = |D_{++}(j\omega_1, j\omega_2)\, D_{+-}(j\omega_1, j\omega_2)|^2. \tag{3.63}$$

10 See Sect. 3.4.2 for details on spectral factorization.

Now, consider the magnitude specification $M(j\omega_1,j\omega_2)$ which we wish to approximate using QP filters. We can note that $M(j\omega_1,j\omega_2)$ satisfies the requirements of a spectral function. We can therefore spectrally factor $M(j\omega_1,j\omega_2)$ as given in (3.63) from which it becomes obvious that *we need both* $++$ *and* $+-$ *QP filters to realize a general magnitude function.* Conversely, $++$ *QP filters can be used to realize only a restricted class of magnitude functions* which factor into min–min and max–max phase terms.

The spectral factorization given in (3.60) is only one of many possibilities. For example, we can factor $S(z_1,z_2)$ into

$$S(z_1,z_2)=D_{\oplus +}(z_1,z_2)\,D_{\ominus -}(z_1,z_2) \quad \text{or} \quad r(m,n)=d_{\oplus +}(m,n)*d_{\ominus -}(m,n)\,, \tag{3.64}$$

where

$$D_{\ominus -}(z_1,z_2)=D_{\oplus +}(z_1^{-1},z_2^{-1}) \tag{3.65}$$

and hence

$$S(j\omega_1,j\omega_2)=|D_{\oplus +}(j\omega_1,j\omega_2)|^2 \tag{3.66}$$

from which it becomes clear that *NSHP recursive filters can be used to realize the general class of positive definite magnitude functions.*

3.4.1 Design of Special Class of NSHP Filters [3.19]

From the preceding discussion on spectral factorization and from [3.1], it follows that a mixed-min phase NSHP array can be decomposed into two QP arrays. For example, we can write $d_{\oplus +}$ as

$$d_{\oplus +}(m,n)=d_{++}(m,n)*d^3_{-+}(m,n) \quad \text{or} \quad D_{\oplus +}(z_1,z_2)=D_{++}(z_1,z_2)D^3_{-+}(z_1,z_2)\,, \tag{3.67}$$

where d^3_{-+} has support in the region $\mathscr{R}^3_{-+}$. Again, d_{++} and d^3_{-+} are of infinite order even if $d_{\oplus +}$ is of finite order and the decomposition is not unique. We can also factor $d_{\oplus +}$ into

$$d_{\oplus +}(m,n)=d_{++}(m,n)*d^1_{-+}(m,n) \quad \text{or} \quad D_{\oplus +}(z_1,z_2)=D_{++}(z_1,z_2)D^1_{-+}(z_1,z_2)\,, \tag{3.68}$$

where d^1_{-+} has support in the region $\mathscr{R}^1_{-+}$.

Equations (3.67, 68) suggest that we can obtain a special class of NSHP filters by writing the denominator polynomial $D_{\oplus +}(z_1,z_2)$ of the transfer function as a product of two QP polynomials as given there and where d_{++} and d^1_{-+} (or d^3_{-+}) are of finite dimensions. Restricting our attention to (3.68), we have

$$H_{\oplus+}(z_1,z_2)=\frac{A_{\oplus+}(z_1,z_2)}{D_{\oplus+}(z_1,z_2)}\equiv\frac{A_{\oplus+}(z_1,z_2)}{D_{++}(z_1,z_2)D^1_{-+}(z_1,z_2)}$$

$$=\frac{\sum_{m=0}^{m_2} a_{\oplus+}(m,0)z_1^{-m}+\sum_{m=-m_2'}^{m_2}\sum_{n=1}^{n_2} a_{\oplus+}(m,n)z_1^{-m}z_2^{-n}}{\left(\sum_{m=0}^{m_1}\sum_{n=0}^{n_1} d_{++}(m,n)z_1^{-m}z_2^{-n}\right)\left(\sum_{m=-m_1'}^{0}\sum_{n=0}^{n_1} d^1_{-+}(m,n)z_1^{-m}z_2^{-n}\right)} \tag{3.69}$$

where

$$D_{++}(z_1,z_2)\neq 0 \quad \text{for} \quad \{|z_1|\geqq 1, |z_2|\geqq 1\} \tag{3.70}$$

and

$$D^1_{-+}(z_1,z_2)\neq 0 \quad \text{for} \quad \{|z_1|\leqq 1, |z_2|\geqq 1\} \tag{3.71a}$$

$$D^1_{-+}(z_1,\infty)\equiv d^1_{-+}(0,0)\,. \tag{3.71b}$$

We now briefly show how to obtain transfer functions, as given in (3.69), where the stability is guaranteed. The approach is based on the 2V network concept introduced in Sect. 3.3.

The polynomial $D_{++}(z_1,z_2)$ is readily obtained using bilinear transformation from the 2V polynomials $Q(s_1.s_2,y_{kl})$ in (3.37) which remains strictly-Hurwitz for all real $\{y_{kl}\}$. The polynomial $Q^1_{-+}(s_1,s_2)$ derived from $D^1_{-+}(z_1,z_2)$ will have the property

$$Q^1_{-+}(s_1,s_2)\neq 0 \quad \text{for} \quad (\mathrm{Re}\{s_1\}\leqq 0, \mathrm{Re}\{s_2\}\geqq 0)\,, \tag{3.72a}$$

$$\mathrm{Q}^1_{-+}(\mathrm{s}_1,1)=\mathrm{k}(1-\mathrm{s}_1)^{m_1'} \qquad k>0\,. \tag{3.72b}$$

A polynomial $Q_{-+}(s_1,s_2,y_{kl})$, having the property given in (3.72a), is obtained by making the transformation $s_1\Rightarrow -s_1$, replacing Y by $Y'=[y'_{kl}]$ (to distinguish it from the Y matrix used before) in the matrix $\hat{Y}_{22}$ of (3.35) and using the resulting matrix in (3.36, 37). Here, of course, the number of diagonal s_1 terms are changed to m_1' with a corresponding change in the dimensions of Y' and $\hat{Y}_{22}$. $Q_{-+}(s_1,s_2,y'_{kl})$ will retain this property for all real values of the coefficients $\{y'_{kl}\}$. Now, using a similar argument as given in Sect. 3.3.1, we can conclude that $Q_{-+}(s_1,s_2=1,y'_{kl})$ will have all its zeros in the right half-plane of s_1. Therefore, by constraining $\{y'_{kl}\}$ properly, we can force $Q_{-+}(s_1,s_2=1,y'_{kl})$ to have the property in (3.72b).

From the above, it becomes clear that we can rewrite $H_{\oplus+}(z_1,z_2)$ as

$$H_{\oplus+}(z_1,z_2)$$
$$=\frac{A_{\oplus+}(z_1,z_2)}{Q\left(\frac{z_1-1}{z_1+1},\frac{z_2-1}{z_2+1},y_{kl}\right)Q_{-+}\left(\frac{z_1-1}{z_1+1},\frac{z_2-1}{z_2+1},y'_{kl}\right)(z_1+1)^{m_1+m_1'}(z_2+1)^{2n_1}}, \tag{3.73}$$

where $Q(s_1, s_2, y_{kl})$ is as given in (3.37) and $Q_{-+}(s_1, s_2, y'_{kl})$ is obtained in a similar way as noted above. We can use any standard optimization technique to obtain a $H_{\oplus +}(z_1, z_2)$ that satisfies the given specifications. The variables of optimization will be the coefficients $\{y_{kl}\}$, $\{y'_{kl}\}$ and $a_{\oplus +}(m, n)$. The $a_{\oplus +}(m, n)$ and $\{y_{kl}\}$ can take any real values and the $\{y'_{kl}\}$ have to be constrained so that (3.72b) is satisfied.

Example 4: A Fan Filter Design

Let us now design a NSHP filter, with $H_{\oplus +}(z_1, z_2)$ as given in (3.69), approximating a fan filter specification. That is, the desired magnitude response is given by

$$M(\mathrm{j}\omega_1, \mathrm{j}\omega_2) = \begin{cases} 1 & |\omega_2| \leqq |\omega_1| \\ 0 & \text{otherwise} \end{cases} \tag{3.74}$$

which is depicted in Fig. 3.11. We set $m_1 = m'_1 = m_2 = m'_2 = n_1 = 2$ and $n_2 = 4$ in (3.69) and reformulated the transfer function as shown in (3.73) to ensure stability. A modified form of the *Fletcher–Powell* algorithm [3.20] is then used to minimize the mean-squared error between the desired response and the

Table 3.4. Coefficients of the fan filter in example 4

$a_{\oplus +} =$

$m \downarrow$ / $\rightarrow n$	0	1	2	3	4
−2	0.0	6.3439	−5.6861	−16.319	10.156
−1	0.0	−21.682	−3.3271	37.461	4.4599
0	−9.8444	41.879	40.580	−11.551	8.2987
1	21.0186	−10.869	−15.241	18.983	−9.0385
2	0.66459	1.7462	−7.5682	−20.891	7.4066

$d_{++} =$

$m \downarrow$ / $\rightarrow n$	0	1	2
0	26.609	−6.6907	−0.21927
1	14.788	−9.8069	−3.7820
2	4.5922	−3.8120	−4.67219

$d^1_{-+} =$

$m \downarrow$ / $\rightarrow n$	0	1	4
−2	0.0	0.29322	−1.16333
−1	0.0	1.83268	−5.68187
0	8.01409	−0.93804	0.98987

Fig. 3.11

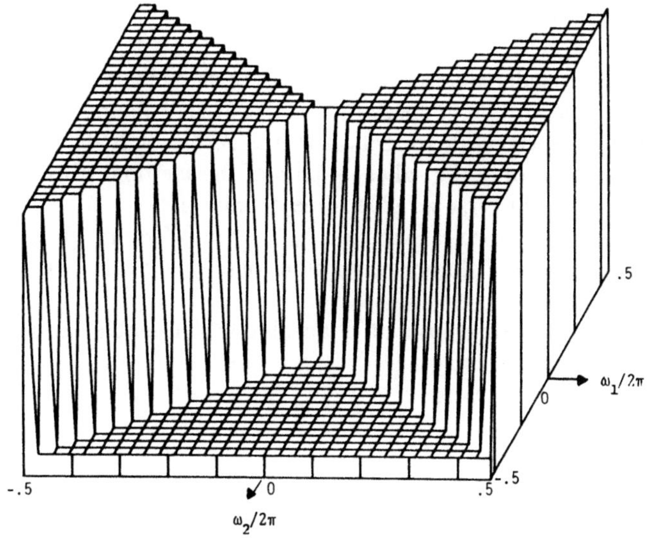

Fig. 3.12

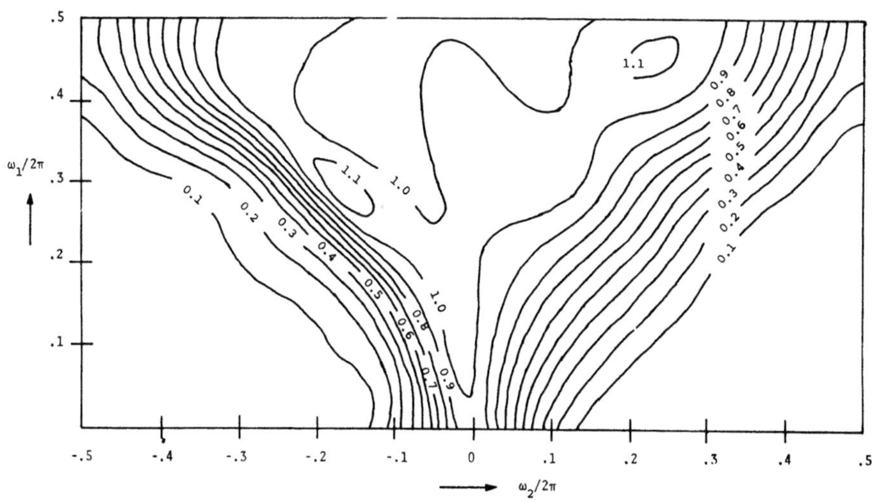

Fig. 3.11. Ideal 90° fan filter specification

Fig. 3.12. Magnitude response of the 2D fan filter in example 4

actual response using $a_{\oplus+}(m,n)$, $\{y_{kl}\}$ and $\{y'_{kl}\}$ as the variables of optimization. (The values of y_{24}, y_{26}, y'_{24}, and y'_{26} were arbitrarily set to zero to reduce the number of variables of optimization.) The result obtained at the end of 200 iterations appears in Table 3.4. In Fig. 3.12, a contour plot of the actual magnitude response $H_{\oplus+}(j\omega_1, j\omega_2)$ is given. As can be observed from this figure, we have obtained a good approximation to a fan filter using this special class of NSHP filters.

3.4.2 NSHP Filter Design Using Spectral Factorization

So far, we have briefly considered spectral factorization as a means to *analyze* QP and NSHP filters and to *obtain* special classes of NSHP filters. Here we study the application of spectral factorization to the *design* of NSHP filters. Again we intend to present only the salient points of the method. (Complete details can be found in [3.1].) First we give some definitions:

i) The decomposition of a spectral function $S(z_1, z_2)$ into factors which are free of zeros or poles in certain distinguished regions of (z_1, z_2) is called *spectral factorization.*

ii) The *2D homomorphic transform* of a absolutely summable sequence $b(m, n)$ (i.e., $b \in l^1$) is given by

$$\mathscr{H} = Z^{-1} \,\mathrm{Ln}\, Z, \tag{3.75}$$

where Z is the 2D z-transform operator and Ln is the pointwise natural logarithm. The inverse z transform is taken on the unit bidisc T^2. Thus the z transform and the inverse are equivalent to the 2D Fourier transform $\mathscr{F}$ and its inverse $\mathscr{F}^{-1}$. Now, $\hat{d}$ as defined above exists and is in l^1 if $D(j\omega_1, j\omega_2)$, the Fourier series of d, is greater than zero. In Fig. 3.13, we have given the block diagram of the $\mathscr{H}$ transform and its inverse $\mathscr{H}^{-1} = Z^{-1} \exp Z$. Note that, in actual numerical implementation, the Fourier transform $\mathscr{F}$ and its inverse $\mathscr{F}^{-1}$ are, respectively, replaced by the discrete Fourier transform (DFT) and the inverse DFT.

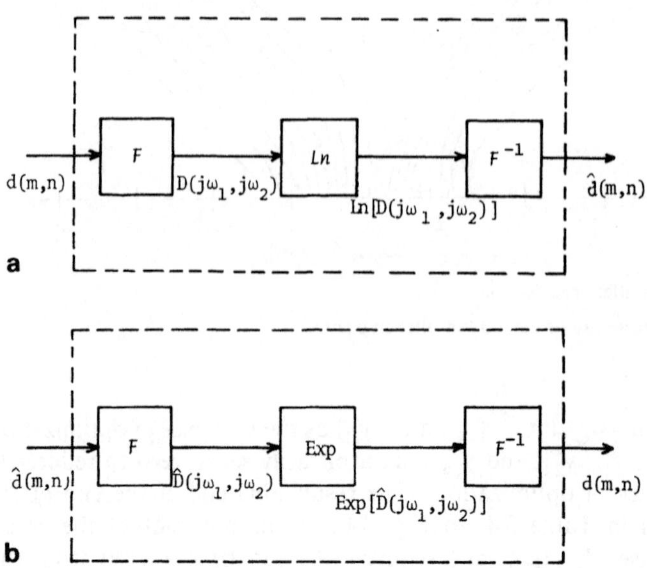

Fig. 3.13. **(a)** Homomorphic transform H; **(b)** inverse homomorphic transform H^{-1}

iii) 2D spectral factorization is determined by a decomposition of the range of $\mathscr{H}$ into admissible regions (possibly overlapping at the boundaries) such as QP, NSHP along with a corresponding set of projection operators whose sum is the identity. The factors are then the inverse transform of the projections onto these regions.

iv) If we project $\hat{d} \in l^1$ onto a sector which is less than or equal to 180° with a projection operator given by P, then $d = \mathscr{H}^{-1}P(\hat{d})$ will have support in this sector (Lemma 1 [3.1]).

v) Let $\hat{d} = \mathscr{H}(d) \in l^1$ and P be a projection operator which projects onto an admissible region. Then $d = \mathscr{H}^{-1}P(\hat{d})$ is recursively computable and stable (Theorem 1 [3.1]).

From the above lemma and theorem, it becomes clear that by sectoring the $\mathscr{H}$ plane into admissible regions (QP, NSHP, etc.) *we can obtain a spectral factorization with factors that are recursively computable and stable.* Thus, we can factor a positive definite function or autocorrelation function $r \in l^1$ into four QP factors or two NSHP factors as given in (3.61) and (3.64), respectively.

Application to Filter Design

The principles of spectral factorization can be applied to 2D recursive filter design since the magnitude function fulfills the requirements of a spectral function. Suppose we are given $M(\mathrm{j}\omega_1, \mathrm{j}\omega_2)$ and wish to approximate it with an all-pole $\oplus+$ NSHP filter transfer function $H_{\oplus+}(z_1, z_2) = 1/D_{\oplus+}(z_1, z_2)$. Then we simply factor $[M(\mathrm{j}\omega_1, \mathrm{j}\omega_2)]^{-1}$ using the 2 factor canonical decomposition as

$$[M(\mathrm{j}\omega_1, \mathrm{j}\omega_2)]^{-1} = |D_{\oplus+}(\mathrm{j}\omega_1, \mathrm{j}\omega_2)\, D_{\ominus-}(\mathrm{j}\omega_1, \mathrm{j}\omega_2)| = |D_{\oplus+}(\mathrm{j}\omega_1, \mathrm{j}\omega_2)|^2$$

to obtain $D_{\oplus+}(z_1, z_2)$ which is mixed-min phase polynomial. That is, referring to Fig. 3.13a, we first obtain $\hat{d}(m, n)$ corresponding to $[M(\mathrm{j}\omega_1, \mathrm{j}\omega_2)]^{-1}$ using DFT and IDFT in the place of Fourier transform and its inverse. Then we project $\hat{d}(m, n)$ onto the NSHP regions $\oplus+$ and $\ominus-$ to obtain $\hat{d}_{\oplus+}(m, n)$ and $\hat{d}_{\ominus-}(m, n)$, respectively. The homomorphic inverse operation, as shown in Fig. 3.13b, is performed on $\hat{d}_{\oplus+}(m, n)$ to obtain $d_{\oplus+}(m, n)$ which is mixed-min phase[11].

Recently, *Ekstrom* et al. [3.21] have incorporated the above spectral factorization into a constrained nonlinear optimization and have proposed a method for the design of stable NSHP filters. (It can also be used for the design of other types of recursive filters.) The method essentially proceeds as follows: Letting $H_{\oplus+}(z_1, z_2) = A_{\mathscr{S}\!f}(z_1, z_2)/D_{\oplus+}(z_1, z_2)$ be the transfer function and $\Phi = [a_{\mathscr{S}\!f}(m, n),\, d_{\oplus+}(m, n)]$ be the parameter vector, a composite error function is formed as

$$J(\Phi) = J_1(\Phi) + J_2(\Phi),$$

11 The factor $d_{\oplus+}(m, n)$ obtained as above will be of infinite order. Truncation and windowing can be used to obtain a polynomial $D_{\oplus+}(z_1, z_2)$ which is of finite order.

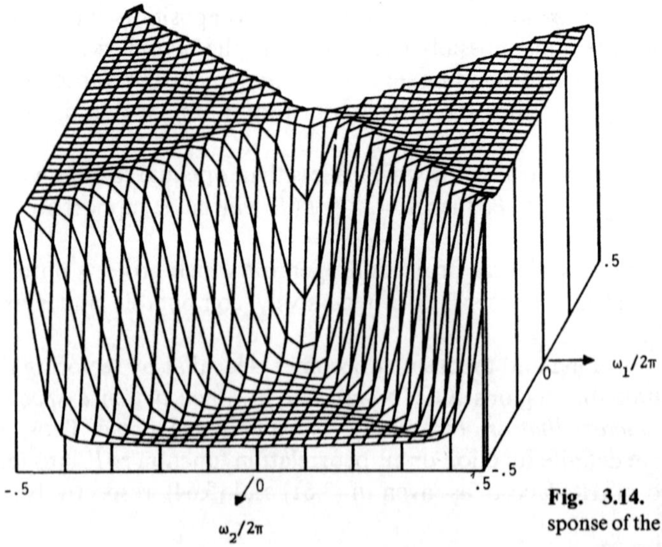

Fig. 3.14. Magnitude-squared response of the 2D fan filter in example 5

where $J_1(\Phi)$ is a suitable error function between the desired frequency response and the actual response and $J_2(\Phi)$, termed as "stability" error function by the authors, is given by

$$J_2(\Phi)=\alpha\sum_m\sum_n[d_{\oplus+}(m,n)-d'_{\oplus+}(m,n)]^p .$$

Here α is a scaling constant and $d'_{\oplus+}(m,n)$ is a *stable mixed-min phase term that is obtained by spectrally factoring* $D_{\oplus+}(j\omega_1, j\omega_2)$. Then, the error function $J(\Phi)$ is minimized using a nonlinear optimization technique. When convergence is achieved, $J_2(\Phi)\to 0$, giving a stable $\oplus+$ transfer function.

Example 5: Fan Filter Design

In Fig. 3.14, we have given the magnitude squared response of a fan filter designed using the above approach. For this example, the transfer function is expressed as $H_{\oplus+}(z_1,z_2)=A_{++}(z_1,z_2)/D_{\oplus+}(z_1,z_2)$. That is, the numerator is restricted to a QP factor. The degree of both A_{++} and $D_{\oplus+}$ is 3 in all directions.

3.4.3 Design of Stable NSHP Filters – A 2V Network Approach

Here, a method developed by the authors [3.22] for the design of NSHP filters is presented. The method has the property that the filters obtained are guaranteed to be stable and can be used to obtain arbitrary-order NSHP filter transfer functions, which for the case of $\oplus +$ filters is given by

$$H_{\oplus +}(z_1, z_2) = \frac{A_{\oplus +}(z_1, z_2)}{D_{\oplus +}(z_1, z_2)}$$
$$= \frac{\sum_{m=0}^{m_2} a_{\oplus +}(m,0) z_1^{-m} + \sum_{m=-m'_2}^{m_2} \sum_{n=1}^{n_2} a_{\oplus +}(m,n) z_1^{-m} z_2^{-n}}{\sum_{m=0}^{m_1} d_{\oplus +}(m,0) z_1^{-m} + \sum_{m=-m'_1}^{m_1} \sum_{n=1}^{n_1} d_{\oplus +}(m,n) z_1^{-m} z_2^{-n}}. \quad (3.76)$$

The motivation for the method has come from the success of the authors in designing stable QP filters using 2V passive network concepts. We considered this design procedure in detail in Sect. 3.3 and as noted there

$$\left\{\begin{array}{l}\text{Denominator of the transfer}\\ \text{function of a stable } ++\\ \text{recursive filter}\end{array}\right\} \overset{\text{BT}}{\Longleftrightarrow} \left\{\begin{array}{l}\text{sum of the numerator and}\\ \text{the denominator of the DPAF}\\ \text{of a 2V reactive network}\end{array}\right\}$$

Stated loosely, there is a one-to-one correspondence between QP recursive filters and 2V reactive networks. This immediately raises the question as to whether there is a continuous system (which is realizable or not) corresponding to NSHP filters in the discrete domain? An answer to this question will be given in the affirmative in the following pages.

The first obvious step is to examine the stability property in terms of the bilinear transformed polynomial $Q(s_1, s_2)$. The properties are given in (3.17) and are reproduced below:

$$D(z_1, z_2) \neq 0 \quad \text{in} \quad \bar{U}^{20} \Rightarrow Q(s_1, s_2) \neq 0 \quad \text{for} \quad (\text{Re}\{s_1\} = 0, \text{Re}\{s_2\} \geqq 0), \quad (3.77\text{a})$$

$$D(z_1, \infty) \neq 0 \quad \text{for} \quad \{|z_1| \geqq 1\} \Rightarrow Q(s_1, 1) = (1 - s_1)^{m'_1} Q_1(s_1), \quad (3.77\text{b})$$

where $Q_1(s_1)$ is 1V strictly-Hurwitz. Now consider the region $\text{Re}\{s_1\} = 0$. We can write this region as

$$(\text{Re}\{s_1\} = 0) = (\text{Re}\{s_1\} \geqq 0) \cap (\text{Re}\{s_1\} \leqq 0 \quad (3.78)$$

or

$$(\text{Re}\{s_1\} = 0) = [(\text{Re}\{s_1\} \geqq 0) \cap (\text{Re}\{s_3\} \geqq 0)]_{s_3 = -s_1}. \quad (3.79)$$

From (3.77a) and (3.79) it becomes clear that $Q(s_1, s_2)$ can be obtained from a 3V polynomial $Q'(s_1, s_2, s_3)$ as

$$Q(s_1, s_2) = Q'(s_1, s_2, s_3)|_{s_3 = -s_1} \quad (3.80)$$

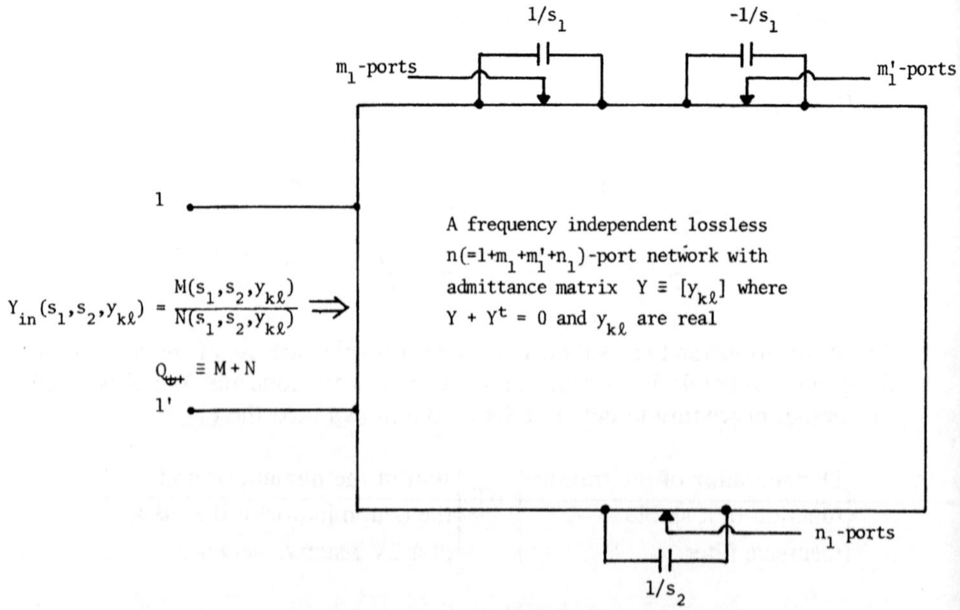

Fig. 3.15. A 2V network with positive and negative capacitors in one variable s_1

and where

$$Q'(s_1, s_2, s_3) \neq 0 \quad \text{for} \quad (\mathrm{Re}\{s_1\} \geqq 0, \mathrm{Re}\{s_2\} \geqq 0, \mathrm{Re}\{s_3\} \geqq 0). \tag{3.81}$$

Equation (3.81) is the definition of a strictly-Hurwitz polynomial (in 3V of course) and a polynomial having this property can be obtained from a 3V reactive network [3.26].

Note, from the preceding statements, that a polynomial $Q_{\oplus +}(s_1, s_2, y_{kl})$ having the property in (3.77a) can be obtained by equating it to the sum of the numerator and the denominator of the DPAF of a 2V network containing both positive and negative reactive elements in s_1 and only positive reactive elements in s_2. One such network is shown in Fig. 3.15 and where $\{y_{kl}\}$ are the only variables. $Q_{\oplus +}(s_1, s_2, y_{kl})$ will retain the property given in (3.77a) for all real $\{y_{kl}\}$. In Table 3.5, we have given the polynomial $Q_{\oplus +}(s_1, s_2, y_{kl})$ as a function of $\{y_{kl}\}$ for $m_1 = m_1' = n_1 = 1, 2$.

Now suppose we set $s_2 = 1$ in the polynomial $Q_{\oplus +}(s_1, s_2, y_{kl})$. From (3.77a) and Fig. 3.15 (replacing each s_2 capacitor by unit valued resistors), it is easy to infer that

$$Q_{\oplus +}(s_1, s_2 = 1, y_{kl}) \neq 0 \quad \text{for} \quad (\mathrm{Re}\{s_1\} = 0). \tag{3.82}$$

Therefore, we can factor $Q_{\oplus +}(s_1, s_2 = 1, y_{kl})$ as

$$Q_{\oplus +}(s_1, s_2 = 1, y_{kl}) = Q_1(s_1)\, Q_2(s_1) = \left(\sum_{i=0}^{m_1} q_{1i} s_1^i\right)\left(\sum_{i=0}^{m_1'} q_{2i} s_1^i\right), \tag{3.83}$$

Table 3.5. The polynomial $Q_{\oplus+}(s_1, s_2, y_{kl})$ used in the design of NSHP filters

m_1	m_1'	n_1	n	$Q_{\oplus+}(s_1, s_2, y_{kl})$ {some of the y_{kl} have been assumed to be zero}
1	1	1	4	$[-s_2-y_{14}^2]s_1^2+[(y_{13}^2-y_{12}^2)s_2+(y_{34}^2-y_{24}^2)]s_1+[y_{23}^2s_2+(y_{12}y_{34}-y_{13}y_{24}+y_{14}y_{23})^2]$
1	1	2	5	$[-s_2^2-y_{14}^2s_2-y_{45}^2]s_1^2+[(y_{13}^2-y_{12}^2)s_2^2+(y_{34}^2+y_{35}^2-y_{24}^2-y_{25}^2)s_2+(y_{13}y_{45}-y_{14}y_{35})^2$ $-(y_{12}y_{45}-y_{14}y_{25})^2]s_1+[y_{23}^2s_2^2+\{(y_{12}y_{34}-y_{13}y_{24}+y_{14}y_{23})^2$ $+(y_{12}y_{35}-y_{13}y_{25})^2\}s_2+(y_{23}y_{45}-y_{24}y_{35}+y_{25}y_{34})^2]$
2	1	2	6	$[-s_2^2-y_{15}^2s_2-y_{56}^2]s_1^3+[(y_{14}^2-y_{12}^2)s_2^2+(y_{45}^2+y_{46}^2-y_{35}^2-y_{36}^2-y_{25}^2-y_{26}^2)s_2$ $+(y_{14}y_{56}-y_{15}y_{46})^2-(y_{15}y_{36})^2-(y_{12}y_{56}-y_{15}y_{26})^2]s_1^2+\{(y_{34}^2-y_{23}^2)s_2^2$ $+[(y_{12}y_{45}-y_{14}y_{25})^2+(y_{12}y_{46}-y_{14}y_{26})^2+(y_{14}y_{35}-y_{15}y_{34})^2$ $+(y_{14}y_{36})^2-(y_{12}y_{36})^2-(y_{12}y_{35}+y_{15}y_{23})^2]s_2+a_1^2+(y_{25}y_{46}-y_{26}y_{45})^2-a_2^2\}s_1$ $+\{(y_{12}y_{34}+y_{14}y_{23})^2s_2^2+[(y_{23}y_{45}+y_{25}y_{34})^2+a_3^2]s_2+(y_{12}a_1+y_{14}a_2-y_{15}a_3)^2\}$ {where $a_1=y_{34}y_{56}-y_{35}y_{46}+y_{36}y_{45}$; $a_2=y_{23}y_{56}-y_{25}y_{36}+y_{26}y_{35}$; $a_3=y_{23}y_{46}+y_{26}y_{34}$}
2	2	2	7	$[s_2^2+y_{16}^2s_2+y_{67}^2]s_1^4+[a_{12}s_2^2+\{a_{10}-a_{11}+y_{36}^2+y_{37}^2-y_{46}^2-y_{47}^2\}s_2$ $+(y_{12}y_{67}-y_{16}y_{27})^2+y_{16}^2(y_{37}^2-y_{57}^2)-(y_{14}y_{67}-y_{16}y_{47})^2]s_1^3+\{(y_{45}^2+y_{23}^2-y_{34}^2)s_2^2$ $+[(y_{12}y_{36}+y_{16}y_{23})^2+y_{37}^2a_{12}+a_4^2+y_{14}^2y_{57}^2-y_{12}^2a_{11}-(y_{12}y_{46}-y_{14}y_{26})^2$ $-(y_{12}y_{47}-y_{14}y_{27})^2-(y_{14}y_{36}-y_{16}y_{34})^2]s_2+[a_1^2-a_5^2-a_6^2-a_7^2+a_8^2$ $-(y_{26}y_{47}-y_{27}y_{46})^2]\}s_1^2+\{[y_{12}^2y_{45}^2-(y_{12}y_{34}+y_{14}y_{23})^2]s_2^2$ $+[a_2^2+a_3^2-a_9^2+y_{45}^2a_{10}-y_{23}^2a_{11}-(y_{23}y_{46}+y_{26}y_{34})^2]s_2$ $+[(y_{12}a_1-y_{14}a_5-y_{16}y_{27}y_{45})^2+(y_{14}a_6+y_{16}a_2)^2-(y_{12}a_7+y_{14}a_8-y_{16}a_9)^2$ $-(y_{12}a_6-y_{16}y_{23}y_{57})^2]\}s_1+\{y_{23}^2y_{45}^2s_2^2+[(y_{12}a_3+y_{23}a_4)^2+(y_{12}a_2+y_{14}y_{23}y_{57})^2]s_2$ $+(y_{23}a_1-y_{26}a_2+y_{27}a_3)^2\}$ where $a_1=y_{45}y_{67}-y_{46}y_{57}+y_{47}y_{56}$; $a_2=y_{34}y_{57}+y_{37}y_{45}$; $a_3=y_{34}y_{56}+y_{36}y_{45}$; $a_4=y_{14}y_{56}+y_{16}y_{45}$; $a_5=-y_{26}y_{57}+y_{27}y_{56}$; $a_6=-y_{36}y_{57}+y_{37}y_{56}$; $a_7=y_{34}y_{67}-y_{36}y_{47}+y_{37}y_{46}$; $a_8=y_{23}y_{67}-y_{26}y_{37}+y_{27}y_{36}$; $a_9=y_{23}y_{47}+y_{27}y_{34}$; $a_{10}=y_{26}^2+y_{27}^2$; $a_{11}=y_{56}^2+y_{57}^2$; $a_{12}=y_{12}^2-y_{14}^2$

where

$$Q_1(s_1)\neq 0 \quad \text{for} \quad (\mathrm{Re}\{s_1\}\geqq 0) \quad \text{and} \quad Q_2(s_1)\neq 0 \quad \text{for} \quad (\mathrm{Re}\{s_1\}\leqq 0) \tag{3.84}$$

and by constraining the values of $\{y_{kl}\}$, we can force $Q_2(s_1)$ [and hence $Q_{\oplus+}(s_1, s_2=1, y_{kl})$] to have m_1' zeros exactly at $s_1=1$.
That is

$$Q_{\oplus+}(s_1, s_2=1, y_{kl})=(1-s_1)^{m_1'}\,Q_1(s_1) \tag{3.85}$$

which is exactly the condition we have in (3.77b). Therefore we can write

$$\left\{\begin{array}{l}\text{Denominator of a}\\ \text{stable } \oplus + \text{ NSHP filter}\\ \text{transfer function}\end{array}\right\} \overset{\text{BT}}{\Longleftrightarrow} \left\{\begin{array}{l} Q_{\oplus +}(s_1, s_2, y_{kl}) = M(s_1, s_2) + N(s_1, s_2)\\ \text{where } M/N = Y_{in} \text{ is the DPAF of a 2V}\\ \text{reactive network with positive and}\\ \text{negative reactive elements in } s_1 \text{ and}\\ \text{positive reactive elements in } s_2 \text{ and}\\ Q_{\oplus +}(s_1, s_2 = 1, y_{kl}) \text{ has } m_1' \text{ zeros at } s_1 = 1\end{array}\right\}$$

The application of the above result to the NSHP filter design problem is straightforward. Similar to the QP filter design we can rewrite $H_{\oplus +}(z_1, z_2)$ as

$$H_{\oplus +}(z_1, z_2) = \frac{A_{\oplus +}(z_1, z_2)}{Q_{\oplus +}\left(\frac{z_1 - 1}{z_1 + 1}, \frac{z_2 - 1}{z_2 + 1}, y_{kl}\right)(z_1 + 1)^{m_1 + m_1'}(z_2 + 1)^{n_1}} \tag{3.86}$$

and use the coefficients $a_{\oplus +}(m, n)$ and $\{y_{kl}\}$ as the variables of an optimization. The values of $\{y_{kl}\}$ have to be so constrained that (3.85) is satisfied. Because of the property of $Q_{\oplus +}(s_1, s_2, y_{kl})$, the stability of $H_{\oplus +}(z_1, z_2)$ will be guaranteed at each iteration of the algorithm and for all real $\{y_{kl}\}$.

Remarks

It is easy to show that $Q_{\oplus +}(s_1, s_2, y_{kl})$ will have the additional properties given in (3.17a). We omit it for the sake of brevity.

$Q_{\oplus +}(s_1, s_2, y_{kl})$ obtained by this method might have a finite number of zeros at $\mathrm{Re}\{s_1\} = \mathrm{Re}\{s_2\} = 0$. The comments given in Sect. 3.3 equally apply here.

We noted that the variables $\{y_{kl}\}$ have to be suitably constrained such that (3.85) is satisfied. This can be achieved as follows [3.23]: let $\Phi = [a_{\oplus +}(m, n), y_{kl}]$ be the parameter vector of length N and $J(\Phi)$ be the error function (for example, a mean-squared error) to be minimized. We can rewrite the constraint in (3.85) as

$$G_k(\Phi) = [\partial^{k-1}\{Q_{\oplus +}(s_1, s_2 = 1, y_{kl})\}/\partial^{k-1}s_1]_{s_1 = 1}; \quad k = 1, 2, \ldots, m_1'. \tag{3.87}$$

Now let $\lambda_k (k = 1 \text{ to } m_1')$ be constants whose values are obtained from the set of equations

$$\sum_{i=1}^{N} \sum_{j=1}^{m_1'} \left(\lambda_j \frac{\partial G_j}{\partial X_i} \frac{\partial G_k}{\partial X_i}\right) = \sum_{i=1}^{N} \left(\frac{\partial G_k}{\partial X_i} \frac{\partial J}{\partial X_i}\right); \quad k = 1, 2, \ldots, m_1', \tag{3.88}$$

where $X_i (i = 1 \text{ to } N)$ denotes the elements of Φ. We can then form a new objective function as

$$J'(\Phi) = J(\Phi) - \sum_{k=1}^{m_1'} \lambda_k G_k + B \sum_{k=1}^{m_1'} G_k^2, \tag{3.89}$$

where B is some positive constant ($\simeq 30$). Then we can minimize $J'(\Phi)$ using a standard unconstrained optimization procedure (*Fletcher-Powell* method [3.20], for example). When convergence is achieved $G_k = 0$ and $J(\Phi) = J'(\Phi)$.

Example 6: Fan Filter Design

Let us consider the approximation of the same fan filter specification given in (3.74) using a second-order (in all directions) transfer function $H_{\oplus+}(z_1, z_2)$. The polynomial $Q_{\oplus+}(s_1, s_2, y_{kl})$ (of degree four in s_1 and two in s_2) given in Table 3.5 is used in (3.86) and the coefficients $a_{\oplus+}$ and $\{y_{kl}\}$ are used as the variables of optimization minimizing a mean-squared error function between the desired response and the actual response. The results obtained at the end of 100 iterations are given in Table 3.6. In Fig. 3.16, an equivalued contour plot of the actual magnitude response is given. Note that we have obtained a good approximation to the fan filter specifications using a second-order transfer function. In fact, the final errors for this example and example 4 are found to be approximately the same although we have used a higher-order transfer function in example 4. This is due to the *restricted nature* of the denominator polynomial of example 4 and is an important advantage of the method described here.

Application to the Design of Symmetric Half-Plane Filters

We now briefly outline the design of the so-called symmetric half-plane filters [3.24] using the above 2V network concepts. The SHP filters considered here are recursively implementable and have transfer functions with denominators

Table 3.6. Coefficients of the fan filter in example 6

$a_{\oplus+} =$

↓ m \ →n	0	1	2
−2	0	−92.407	−76.155
−1	0	59.562	42.494
0	12.458	41.972	8.5389
1	7.6765	−13.089	−6.5886
2	2.4377	−21.683	18.215

$d_{\oplus+} =$

↓ m \ →n	0	1	2
−2	0.0	2.6549	−6.1802
−1	0.0	−202.82	74.401
0	367.78	−400.55	184.30
1	358.90	−502.10	209.10
2	120.69	−246.41	100.44

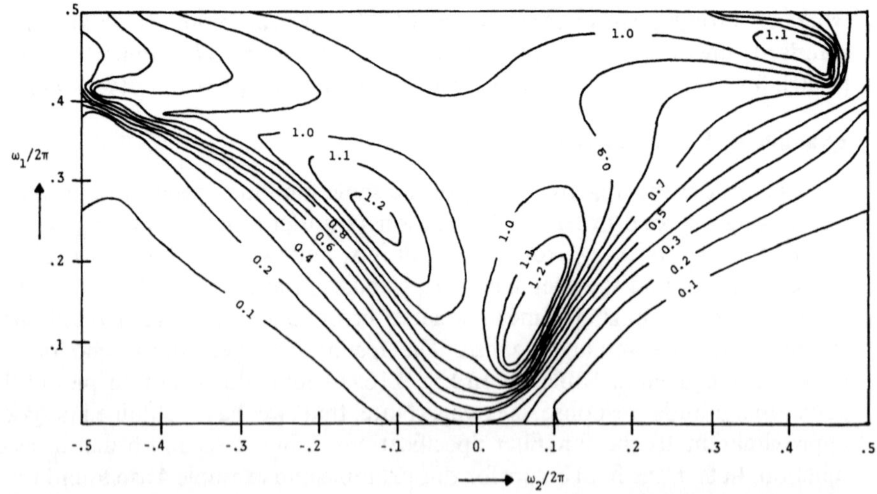

Fig. 3.16. Magnitude response of the 2D fan filter in example 6

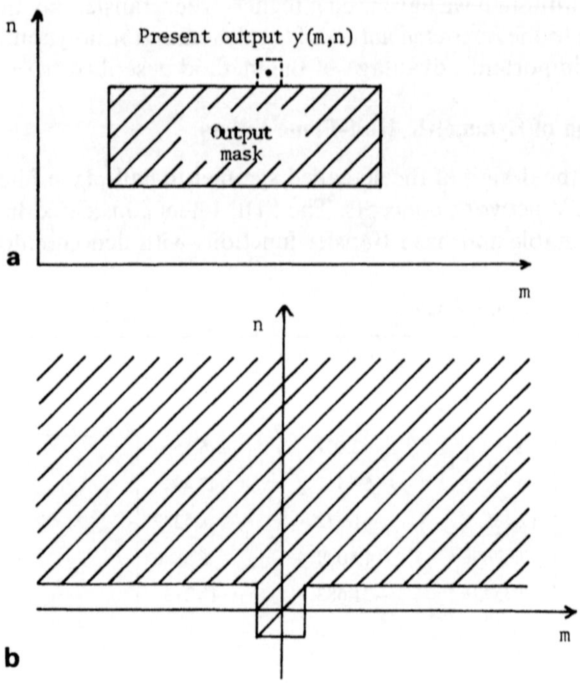

Fig. 3.17. **(a)** 2D recursion using a SHP filter; **(b)** region for its impulse response

of the form

$$D(z_1, z_2) = 1 + \sum_{m=-m'_1}^{m_1} \sum_{n=1}^{n_1} d(m,n) z_1^{-m} z_2^{-n}. \tag{3.90}$$

In Fig. 3.17a, we have shown a 2D recursion using a SHP filter. Comparing this figure with Fig. 3.3b, it can be noted that a SHP filter differs from that of a

NSHP filter by the way the output mask is defined. A SHP filter omits all of the nth row when calculating the output. That is each row of output of a SHP filter depends only on previous rows of output. Therefore all the outputs in a given row can be computed in parallel. *This factor would be of significance when one consideres the implementation of 2D filtering in real time.* Note that the impulse response $h(m,n)$ of a SHP filter has support in the region $\mathscr{R}'_{\oplus +} = \mathscr{R}^1_{++} \cup \mathrm{R}^3_{-+}$ as shown in Fig. 3.17b. That is, $h(m,0) = $ constant for $m=0$ and zero otherwise.

Now, the polynomial $D(z_1, z_2)$ has to be mixed-min phase for the filter to be stable. This is the same as the condition (3.17a) of a NSHP filter. Also from (3.90) it follows that

$$D(z_1, \infty) = 1 \tag{3.91}$$

and the bilinear transformed polynomial $Q(s_1, s_2)$ corresponding to $D(z_1, z_2)$ will have the properties

$$Q(s_1, s_2) \neq 0 \quad \text{for} \quad (\mathrm{Re}\{s_1\} = 0, \mathrm{Re}\{s_2\} \geqq 0), \tag{3.92a}$$

$$Q(s_1, 1) = k(1 - s_1)^{m'_1}(1 + s_1)^{m_1}; \qquad k > 0. \tag{3.92b}$$

Comparing (3.92b) with (3.83, 84) it becomes clear that we can obtain a polynomial $Q(s_1, s_2)$ satisfying (3.92a, b) from the 2V network of Fig. 3.15. Therefore we can use the polynomial $Q_{\oplus +}(s_1, s_2, y_{kl})$ as defined in the figure to obtain stable SHP filters.

3.5 Summary

Techniques for the design of two-dimensional recursive filters are considered. A unified treatment of quarter-plane and half-plane recursive filters is made and many of the recent results on their stability, implementation and design are presented. The earlier work by *Read* et al. [3.25] in an earlier volume of this series has been kept in mind during the writing of this chapter. The results presented there are completely omitted and only the results that appeared after the publication of their work are considered here. In all, three algorithms for the design of QP filters and three algorithms for the design of NSHP filters are presented. The first two algorithms described in Sect. 3.2, for the design of QP filters are restricted to special types of filters and can only give *suboptimum* filters. The third algorithm descirbed in Sect. 3.3, can be used to design *stable* QP filters of *arbitrary order* and has the potential to yield *optimum* filters.

The QP filter algorithms presented here are the straightforward evolution (at least in our view) of three different methods of designing QP filters. The *Maria* and *Fahmy* method represents an earlier (and perhaps the first attempt) to design QP filters using a direct approach. This resulted in a technique in which stability of the filter was not assured and which can give only limited types of transfer functions. In the second method, *SID-Ahmed* and *Julien* were aware that stability is a major problem in the design procedure and overcame this problem using transformation of variables. However, this method is also restricted to special types of transfer functions since it is not possible to apply variable transformation directly to transfer functions of arbitrary order. In the third method, the authors solved the general stability problem using 2V reactive network concepts *and* arrived at a technique for the design of QP filters of arbitrary order (filters of order up to seven in each variable have been designed using this approach [3.14]). Hence this approach can be considered to have *completely solved* the QP filter design problem.

In choosing the algorithms presented here for the design of NSHP filters (Sect. 3.4), we did not face much of a problem since the interest in this class of filters started only a couple of years ago and hence the number of algorithms available is limited. It is quite probable that this number will increase rapidly in the near future.

We have not mentioned different structures for implementing recursive filters. However, the first two algorithms presented for QP filters in Sect. 3.2, are inherently restricted to *cascade* structures and, in fact, the authors have used cascaded forms since they have better sensitivity performance than direct form structures. In the third algorithm presented in Sect. 3.3, the authors have given a solution to the general *approximation problem* leaving aside the question of implementation (it should be pointed out that this third algorithm includes the previous two algorithms as special cases if we restrict the transfer functions to second order). An important feature of the authors' approach is that it leads to *all possible stable transfer functions* whereas a priori restriction to cascaded structures eliminates this feature. Also, as is well known in the cases of 1D recursive filters and RC active filter implementations, *cascaded structures* are *not* the best in terms of sensitivity considerations. The *ladder structure* is extremely well behaved in the case of 1D filters and it should be noted that the insensitivity property of ladder filters is not based on factoring the transfer function but structure-independent properties such as maximum power transfer in the passband. Hence, this simply demonstrates the need to *consider separately* the implementation problem and to obtain 2D *structures* having all the desirable properties. It is needless to say that much work has to be done in this direction. For example, the possible synthesis of 2D structures that retain the maximum power transfer property and retain excellent passband insensitivity to the multiplier coefficients?

Acknowledgement. The work reported here was performed at the Department of Electrical Engineering, University of Calgary, Canada and was made possible by a graduate assistantship to one of the authors (*P. A. Ramamoorthy*) and a research grant from the Natural Science and Engineering Research Council of Canada (No. A 7997). The authors are particularly grateful to Prof. *M. M. Fahmy* of the Queen's University, Kingston, Canada, for providing the design example 1 and Fig. 3.4 and to Prof. *J. W. Woods* of Rensselaer Polytechnic Institute, Troy. New York and co-workers, for providing a preprint of their paper [3.21] and Figs. 3.11 and 14.

References

3.1 M.P.Ekstrom, J.W.Woods: IEEE Trans. ASSP-**24**, 115 (1976)
3.2 D.Goodman: IEEE Trans. CAS-**24**, 209 (1977)
3.3 W.F.Osgood: *Topics in the Theory of Functions of Several Complex Variables* (Dover, New York 1966)
3.4 D.Goodman: Proc. IEEE **66**, 796 (1978)
3.5 H.G.Ansell: IEEE Trans. CT-**11**, 214 (1964)
3.6 T.S.Huang: IEEE Trans. AU-**20**, 158 (1972)
3.7 J.L.Shanks, S.Treitel, J.H.Justice: IEEE Trans. AU-**20**, 115 (1972)
3.8 H.Ozakik T.Kasami: IRE Trans. CT-**7**, 251 (1960)
3.9 T.Koga: IEEE Trans. CT-**13**, 31 (1966)
6.10 G.A.Maria, M.M.Fahmy: IEEE Trans. ASSP-**22**, 15 (1974)
3.11 N.A.Pendergrass: IEEE Trans. ASSP-**23**, 389 (1975)
3.12 M.A.Sid-Ahmed, G.A.Julien: "Frequency Domain Design of a Class of Stable Two-Dimensional Recursive Filters", preprint, University of Windsor, Windsor, Ontario, Canada (June, 1974)
3.13 D.C.Youla: Proc. PIB Symp. Generalized Networks, 289 (1966)
3.14 P.A.Ramamoorthy: "Design Techniques for Stable Single- and Multi-Dimensional Digital Filters", Ph.D. Thesis, Dept. of Elect. Eng., University of Calgary (April 1977)
3.15 P.A.Ramamoorthy, L.T.Bruton: Intl. J. Circuit Theory and Applications, **7**, 229 (1979)
3.16 T.N.Rao: Bell Syst. Tech. J. **48**, 163 (1969)
3.17 R.Fletcher: "Fortran Subroutine for Minimization by Quasi-Newton Methods", Rpt. R 7125 AERE, Harwell, England (1971)
3.18 S.A.H.Aly, M.M.Fahmy: IEEE Trans. CAS-**25**, 908 (1978)
3.19 P.A.Ramamoorthy, L.T.Bruton: "Design of Stable 2D Nonsymmetric Half-Plane Recursive Filters", Proc. 23rd Midwest Symp. on Circuits and Systems (1980)
3.20 R.Fletcher, M.J.D.Powell: Computer J. **6**, 163 (1963)
3.21 M.P.Ekstrom, R.E.Twogood, J.W.Woods: "Two-Dimensional Recursive Filter Design: A Special Factorization Approach", preprint, Lawrence Livermore Laboratory, UCRL, CA (Jan. 1979), also in IEEE Trans. ASSP-28, **16** (1980)
3.22 P.A.Ramamoorthy, L.T.Bruton: Proc. 1979 IEEE Intern. Conf. ASSP (April, 1979) p. 40
3.23 P.C.Harloff, J.D.Buys: Computer J. **13**, 178 (1970)
3.24 J.Murray: 20th Midwest Symp. on Circuits and Systems (1977) p. 434
3.25 R.R.Read, J.L.Shanks, S.Treitel: "Two-Dimensional Recursive Filtering", in *Picture Processing and Digital Filterings*, 2nd ed., ed. by T.S.Huang, Topics in Applied Physics, Vol. 6 (Springer, Berlin, Heidelberg, New York 1979) pp. 131–176
3.26 N.K.Bose, R.W.Newcomb: Intl. J. Electron. **36**, 417 (1974)

4. Stability of General Two-Dimensional Recursive Filters

B. T. O'Connor and T. S. Huang

With 18 Figures

In recent years digital signal processing has grown quickly and its application has spread from one-dimensional to two-dimensional to n-dimensional data. There exist many physical phenomena which are inherently multidimensional. For instance, picture data such as satellite photographs, x-rays, electronmicrographs, and infrared pictures form probably the largest subset of the class of two-dimensional signals. Furthermore, other multidimensional signals such as seismic, magnetic, gravitational, and hydrophonic data are being processed more and more by digital computers. In this chapter only two-dimensional data processing will be considered. However, once the transition from one to two dimensions is understood, the transition required to extend signal processing methods to n-dimensions is usually trivial. On the other hand, the extension of one-dimensional techniques to two dimensions is in many cases very difficult. In particular, the stability properties of two-dimensional recursive filters are many times more complicated than those of one-dimensional recursive filters. This is due mainly to the fact that the Fundamental Theorem of Algebra does not extend to two-dimensional polynomials. In this chapter our main concern will be with several problems in two-dimensional recursive filtering. In particular, stability theorems and efficient stability tests will be developed extensively. However, before this discussion a quick review of digital signal processing fundamentals will be given in order to place recursive filtering in perspective.

4.1 Background

The main purpose of digital signal processing techniques is to manipulate data in a controlled manner. In most cases this manipulation can be described as a transformation of an input sequence into an output sequence. The equation

$$y(m,n)=L[x(m.n)] \tag{4.1}$$

describes this on two-dimensional data where $x(m,n)$ and $y(m,n)$ are the respective input and output sequences and L is an operator or system which maps input into output.

An important class of operators are the linear shift-invariant systems which are characterized by the following relations:

$$L[ax_1(m,n)+bx_2(m,n)]=a\cdot L[x_1(m,n)]+b\cdot L[x_2(m,n)] \tag{4.2}$$
$$L[x(m-m_0,n-n_0)]=y(m-m_0,n-n_0)\,,$$

where a and b are constants and m_0 and n_0 integers. Linear shift-invariant (LSI) systems are widely used because they can be implemented easily. In particular, the input–output relation is represented by the convolutional sum

$$y(m,n)=\sum x(k,l)h(m-k,n-l)\,, \tag{4.3}$$

where (k,l) takes on all integer ordered pairs and $h(m,n)=L[\delta(m,n)]$ where $\delta(m,n)$ is the unit impulse [that is, $\delta(m,n)=1$ if $m=n=0$ and zero elsewhere] and $h(m,n)$ is called the impulse response. The convolutional sum can be employed as a computational realization of an LSI system. Furthermore, $w^m z^n$ are eigenfunctions [4.1] of L so $L[w^m z^n]=H(w,z)w^m z^n$ where

$$H(w,z)=\sum h(m,n)w^{-m}z^{-n}$$

is called the system function. $H(w,z)$ is also called the two-dimensional z transform of $h(m,n)$ and is defined for all (w,z) where the power series converges. Given $H(w,z)$ and a region of convergence, $h(m,n)$ can be found from the following inversion integral:

$$h(m,n)=\frac{1}{(2\pi j)^2}\int\limits_{c_1}\int\limits_{c_2} H(w,z)w^{m-1}z^{n-1}dwdz\,. \tag{4.4}$$

Here c_1 and c_2 are suitable closed contours in a region of convergence of $H(w,z)$ [4.2].

Much of the literature on two-dimensional recursive filters, e.g. [4.3–6], employs a slightly different definition for the system function. Here

$$H(w,z)=\sum h(m,n)w^m z^n \tag{4.5}$$

and for convenience we will adopt this convention for both the one- and two-dimensional z transforms.

If $\sum\limits_{(m,n)}|x(m,n)|<\infty$, that is, $x(m,n)$ is absolutely summable or equivalently, $x\in l^1$ then its Fourier transform exists and the following equations hold:

$$X(u,v)=\sum x(m,n)\exp(-\mathrm{j}(um+vn))\,, \tag{4.6a}$$

$$x(m,n)=\frac{1}{4\pi^2}\int\limits_{-\pi}^{\pi}\int\limits_{-\pi}^{\pi} X(u,v)\exp(\mathrm{j}um+\mathrm{j}vn)dudv\,, \tag{4.6b}$$

and

$$y(m,n)=L[x(m,n)]=\frac{1}{4\pi^2}\int_{-\pi}^{\pi}\int_{-\pi}^{\pi}X(u,v)L[\exp(jum+jvn)]du\,dv\,, \tag{4.7a}$$

where

$$H(u,v)=L[\exp(jum+jvn)] \tag{4.7b}$$

which is the system function evaluated at $w=\exp[-ju]$ and $z=\exp[-jv]$. For convenience, $H(u,v)$ is used instead of the correct $H(e^{-ju},e^{-jv})$. Moreover, $H(u,v)$ is equal to the Fourier transform of $h(m,n)$ if $h\in l^1$ and is called the frequency response of the system. Therefore, if $h\in l^1$ the system can be completely described by $h(m,n)$, $H(u,v)$, or $H(w,z)$. The condition $h\in l^1$ is not very restrictive because it is both a necessary and sufficient condition for BIBO stability. Here, stability means that a bounded input produces a bounded output (BIBO).

An important problem in linear system theory is the approximation of a stable operator L. The major reason for approximating L is to find an operator which "acts" like L while at the same time being computationally easier to apply to the input data $x(m,n)$. For example, the operator may have an infinite extent impulse response which makes it difficult if not impossible in some cases to implement the operator via the convolution sum given in (4.3). Moreover, as will be discussed later, only a small class of operators with infinite extent impulse responses can be implemented by means of recursive equations. Secondly, an operator may be specified by its frequency response $H(u,v)$. It may happen that if (4.7a) is used to apply L or computationally realize L on an input array then the integral may not have a closed form solution. Numerical techniques such as the fast Fourier transform (FFT) can be employed to approximate this integral, however this, in effect, is a method of specifying an approximate operator.

The approximation problem can be approached by several methods. One procedure approximates the system function by a polynomial of finite extent. This allows the system to be practically implemented by either (4.3) or (4.7a). Furthermore, since the impulse response has finite support, the stability of the approximated system is guaranteed. These approximations are usually referred to as finite impulse response (FIR) or nonrecursive methods.

Another method, which will be studied in this chapter, approximates the system function or frequency response as a ratio of finite support polynomials A and B so

$$\tilde{H}(w,z)=A(w,z)/B(w,z)\,, \tag{4.8}$$

where $A(w,z)=\sum a(m,n)w^m z^n$ and $B(w,z)=\sum b(m,n)w^m z^n$. Usually, this approximation produces an impulse response of infinite extent and a computational algorithm which implements this is referred to as an infinite impulse response filter (IIR). For this form to have any practical value it must suggest an implementable computational algorithm which can relate the input and output arrays. Two-dimensional recursive filters can be derived from this approach. The next sections will discuss in detail both the implementation and stability of two-dimensional recursive filters.

4.2 Introductary Comments

Two-dimensional recursive digital filters have generated much interest lately. They have the potential of saving computer time and memory. In virtually all developments of recursive filtering one starts with a two-dimensional system function similar to (4.8) from which a two-dimensional difference equation is obtained. Then a particular solution of this equation which is acquired by an iterative procedure is considered. This iterative procedure represents the two-dimensional filter and also acts as a computational realization of the system. However, this method of developing recursive filters is ambiguous. In general, a difference equation can have many solutions which can be acquired by iterative procedures [4.1, 3]. Researches have tried to avoid this ambiguity by considering only the causal recursive formula. However, this approach is very limiting because in two dimensions causality has little importance. The important point is whether the iterative procedure can be utilized to consistently calculate any output from a given set of initial conditions.

Example 1 (E1)

The following example illustrates the above approach to recursive filtering. Let

$$A(w,z)=\sum_{m=0}^{M}\sum_{n=0}^{N}a(m,n)w^m z^n$$

$$B(w,z)=\sum_{m=0}^{M'}\sum_{n=0}^{N'}b(m,n)w^m z^n\ ,$$

where $a(m,n)$ and $b(m,n)$ are constants and let $X(w,z)$ and $Y(w,z)$ be the z transform of the respective input and output arrays $x(m,n)$ and $y(m,n)$. Therefore, by the definition of the system function we have

$$\frac{Y(w,z)}{X(w,z)}=\frac{A(w,z)}{B(w,z)}\ . \tag{E1.1}$$

After cross-multiplying and inverse z transforming the following two-dimensional difference equation is obtained:

$$\sum_{m=0}^{M'}\sum_{n=0}^{N'}b(m,n)y(r-m,s-n)$$
$$=\sum_{m=0}^{M}\sum_{n=0}^{N}a(m,n)x(r-m,s-n)\ . \tag{E1.2}$$

We can express $y(r,s)$ in terms of the rest [if $b(0,0)\neq 0$] as

$$y(r,s)=\frac{1}{b(0,0)}\left\{\sum_{m=0}^{M}\sum_{n=0}^{N}a(m,n)x(r-m,s-n)\right.$$
$$\left.-\sum_{\substack{m=0\\(m,n)}}^{M'}\sum_{\substack{n=0\\ \neq(0,0)}}^{N'}b(m,n)y(r-m,s-n)\right\} \tag{E1.3}$$

and thus obtain a recursive filter that can be applied in the $(+r,+s)$ direction; that is, a particular solution to the difference equation (E1.2) can be found by appropriately iterating (E1.3) in the $(+r,+s)$ direction. What distinguishes (E1.3) from the difference equation (E1.2) is that it is an explicit procedure for computing the output $y(r,s)$ and for appropriate initial conditions it can be utilized to calculated any $y(r,s)$ for $r\geqq 0$, $s\geqq 0$ in a recursive manner.

Equation (E1.2) can also be solved for $y(M'-r,N'-s)$, $y(r,N'-s)$, or $y(M'-r,s)$ if $b(M',N')$, $b(0,N')$, and $b(M',0)$ are, respectively, nonzero giving recursive equations which recurse, respectively, in the $(-r,-s)$, $(r,-s)$, and $(-r,s)$ directions. Each of these recursive equations can be used to obtain a particular solution to difference equation (E1.2). Therefore, (E1.1) actually represents a number of recursive filters. In fact, (E1.1) could possibly give recursive equations which recurse in directions other than the four given above. Hence, it appears that the system function description of recursive equations is not a good method.

For these reasons in this chapter we shall define recursive filters from a different point of view. Then, we will present a number of stability theorems for general recursive filters. Among these is a theorem which relates the stability of any digital filter to its two-dimensional phase function. Furthermore, it will be shown that any general recursive filter can be mapped into a first-quadrant filter. Also, a number of stability tests will be developed and evaluated. Included in these are several efficient algorithms which utilize the FFT algorithm. Finally, several examples of determining the stability of general two-dimensional recursive filters will be presented.

4.3 General Recursive Filters

Linear constant coefficient recursive equations can be utilized to implement a class of linear shift-invariant (LSI) systems on many input arrays. The form that will be considered is

$$o(m,n)=\sum_{(r,s)\in\alpha}a(r,s)i(m-r,n-s)$$
$$-\sum_{(k,l)\in\beta-(0,0)}b(k,l)o(m-k,n-l)\,, \tag{4.9}$$

where $a(r,s)$ and $b(k,l)$ are real finite extent arrays with respective lattice supports of α and β and $i(m,n)$ and $o(m,n)$ are the respective input and output arrays. Note that support of an array is defined as the set of all ordered integer pairs where the array is not zero. Furthermore, we will assume that $b(0,0)=1$ and that β is contained in a lattice sector with vertex $(0,0)$ of angle less than π. This latter condition means that β, the support of $b(k,l)$, is shaped so that there exists a half plane of lattice points defined by a line passing through the origin such that the half plane intersects β only at $(0,0)$. These conditions guarantee that for all finite extent input arrays and a class of infinite extent inputs, (4.9) can be solved by incrementing the values of the indices (m,n) in such a fashion that all values of the output can be computed in turn from a given set of initial conditions [4.1, 2]. In other words, the stated conditions imply that (4.9) is recursible for many inputs. For the moment we will assume that the input array has finite support and hence (4.9) will always be recursible. If this is the case then it is possible to compute $o(m,n)$ iteratively by using the right-hand side of (4.9) to obtain new output values. We plan to interpret (4.9) in this manner, that is, as a recursive formula rather than a rewritten difference equation which can be derived from the system function. However, it is important to note that the solution obtained by iterating (4.9) is a particular solution of the two-dimensional difference equation that (4.9) also represents. It may be helpful for the reader to view the equal sign in (4.9) as an assignment of the calculated value from the right side to the appropriate output value on the left-hand side. Several examples will be given later showing how (4.9) is applied. Since each output point depends on the same previous output points, initial conditions and input points, (4.9) represents a unique system. In fact, if zero initial conditions are assumed (4.9) is a computational realization of an LSI system with the following impulse response:

$$h(m,n)=a(m,n)-\sum_{(r,s)\in\beta-(0,0)} b(r,s)h(m-r,n-s)\,. \tag{4.10}$$

Again, this will be viewed as a recursive formula rather than a rewritten difference equation. Therefore, for the case of zero initial conditions, applying (4.9) to an input $x(m,n)$ is equivalent to convolving it with the array $h(m,n)$ obtained from (4.10).

If $a(m,n)$ and $b(m,n)$ are first-quadrant arrays, that is, both $a(m,n)$ and $b(m,n)$ are zero for any negative values of m or n, then (4.9) represents a recursive filter which recurses in the $(+m,+n)$ direction. In other words, the output must be computed using previously computed values which are to the left or below the present output point. The stability and the design of first-quadrant quarter-plane recusion filters has been extensively studied in the literature, e.g. [4.2–4, 7]. If $a(m,n)$ and $b(m,n)$ are second-quadrant arrays, that is, both $a(m,n)$ and $b(m,n)$ have supports contained in the second quadrant, then (4.9) recurses in the $(-m,n)$ direction. Third- and fourth-quadrant filters can be defined accordingly. Since the impulse response, as defined by (4.10), of any of these

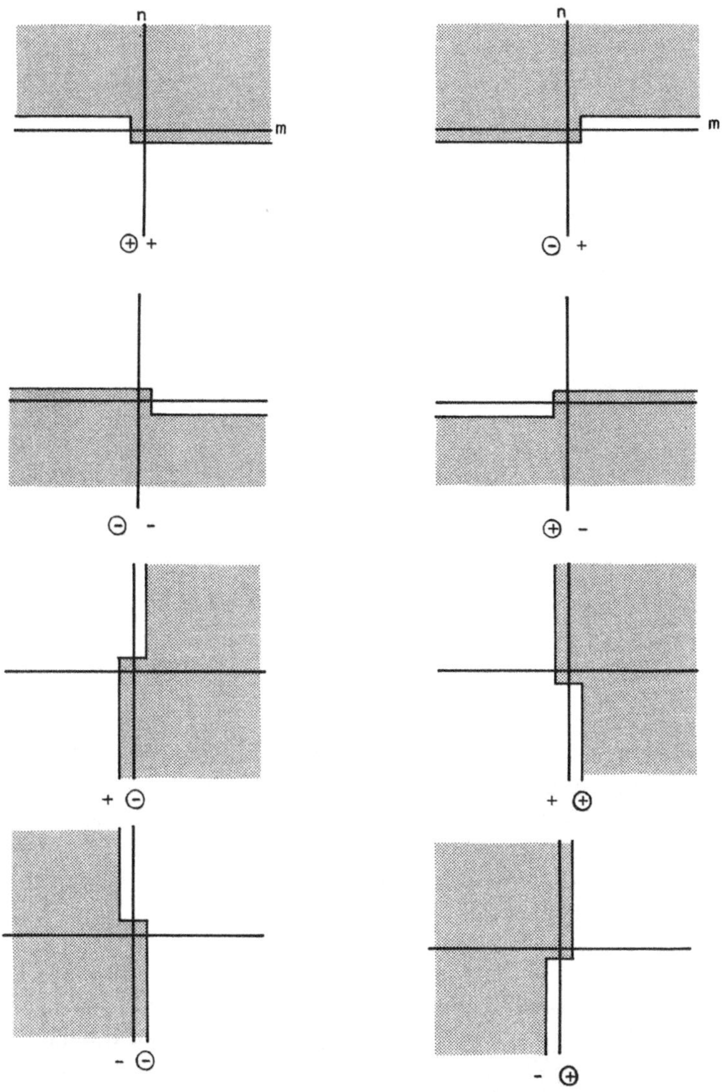

Fig. 4.1 Regions of support for the eight classes of nonsymmetric half-plane filters

filters is supported in one of the quadrants of the (m, n) index plane these filters form a class called quarter-plane recursive filters. A more general class of quarter-plane recursive filters can be defined by removing the restriction that the support of $a(m, n)$ lie in the same quadrant as that of $b(m, n)$. For this more general class the only restriction on $a(m, n)$ is that it have finite support. Hence, in this case, if $b(m, n)$ is a quarter-plane array the filter will recursively realize an impulse response which can be shifted in such a fashion to be contained in the

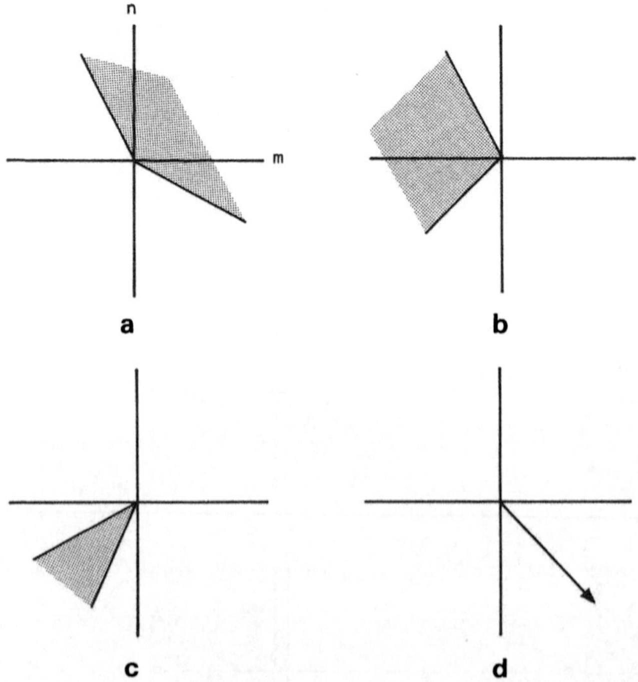

Fig. 4.2a–d. General recursive filter supports

quadrant that $b(m,n)$ lies in. It should be clear from studying (4.9) that the only effect $a(m,n)$ has is to modify the input $i(m,n)$ to the recursive filter. Therefore we will adopt the convention that the $b(m,n)$ array completely defines the filter type. For convenience, notation from [4.8] will be employed to represent various classes of filters. A first-quadrant filter will be said to belong to the $(++)$ class, second-quadrant $(-+)$, third-quadrant $(--)$, and fourth-quadrant $(+-)$.

Equation (4.9) also represents any type of nonsymmetric half-plane filter considered in [4.1, 8]. Figure 4.1 illustrates the eight regions in the (m,n) plane where $b(m,n)$ should be supported in order to be a member of the family of nonsymmetric half-plane filters. Notice that the $(++)$ filter class belongs to both the $(\oplus+)$ and $(+\oplus)$ classes. Furthermore, a more general class of recursive filters can be defined using (4.9) by allowing $b(m,n)$ to be supported in any sector with vertex $(0,0)$ of angle less than π. Figure 4.2 illustrates several regions of support. In case (a) the filters does not belong to any class of nonsymmetrical half-plane filters while example (b) belongs to both the $(\ominus+)$ $(-\ominus)$ classes. Example (c) belongs to the $(--)$ class while example (d) depicts a recursive filter whose support is located on a ray starting from the origin.

At first glance it may seem that an even more general formulation can be given by removing the restrictions on β. However, if β is not a subset of a lattice sector of angle less than π then both (4.9, 10) become inconsistent. No matter how the filtering is performed, numerical values for a number of output points must be assumed before they are calculated. In other words, (4.9, 10) will not be recursively computable. Example 2 illustrates this effect.

Example 2 (E2)

Let $a(m,n)=\delta(m,n)$; $i(m,n)=\delta(m,n)$; and $b(0,0)=1$, $b(1,0)=-e$, $b(0,1)=-f$, $b(-1,-1)=-g$, and zero otherwise. Notice that $b(m,n)$ does not satisfy the constraints of (4.9). Equation (4.9) becomes

$$h(m,n)=\delta(m,n)+e\cdot h(m-1,n)+f\cdot h(m,n-1)+g\cdot h(m+1,n+1)\,. \qquad \text{(E2.1)}$$

Zero initial conditions will be assumed wherever appropriate. In order to start computing (E2.1) assume $h(-1,0)=h(1,1)=0$. Furthermore, $h(1,0)$ and $h(0,1)$ can be computed to give

$$h(1,0)=e\cdot h(0,0)=e$$

$$h(0,1)=f\cdot h(0,0)=f$$

and

$$h(1,1)=eh(0,1)+fh(1,0)=2ef\,.$$

However, this is inconsistent with the assumption that $h(1,1)=0$ unless e or f or g is zero. Therefore, $h(1,1)$ cannot be assumed to be zero. However, any other value assigned to $h(1,1)$ will always lead to an inconsistency somewhere unless one of the coefficients is zero. Therefore, (E2.1) is not a well-defined recursive equation.

For the rest of this chapter, it will be assumed that (4.9) is a computational implementation of a LSI system. In order for superposition to hold this assumption implies that zero initial conditions for $o(m,n)$ must be assumed. Equation (4.10) is a computational procedure which computes the impulse response of the LSI system of which (4.9) implements. In order to solve for $h(m,n)$ it is assumed that zero initial conditions are assigned to values of $h(m,n)$ wherever consistent. As long as β, the support of $b(m,n)$, satisfies the conditions given with (4.9), consistent zero initial conditions always exist for (4.10). However, if the input array in (4.9) has infinite support, problems may arise in applying (4.9) to it, even though all the other conditions are met, which are

- $a(m,n)$ and $b(m,n)$ are real with finite extent
- $b(0.0)=1$
- β, the support of $b(m,n)$, is contained in a lattice sector of vertex $(0,0)$ of angle less than π.

The following example illustrates this problem.

Example 3 (E3)

Assume $a(m,n)=\delta(m,n)b(m,n)=\delta(m,n)-e\cdot\delta(m-1,n)-f\cdot\delta(m,n-1)$ [that is, $b(0,0)=1$, $b(1,0)=-e\neq 0$, $b(0,1)=-f\neq 0$, and zero elsewhere] and $i_1(m,n)$ is zero for $m<0$ and nonzero if $m\geqq 0$. Equation (4.4) or (4.10) can be employed (see [4.2]) to compute the impulse response of the recursive filter as

$$h(m,n)=\begin{cases} e^m f^n \dfrac{m!n!}{(m+n)!} & \text{for } m\geqq 0,\, n\geqq 0 \\ 0 & \text{otherwise}. \end{cases} \tag{E3.1}$$

If e and f are nonzero the impulse response is then nonzero throughout the first quadrant of (m,n) plane.

The recursive equation defined by the given arrays is

$$o(m,n)=i_1(m,n)+e\cdot o(m-1,n)+f\cdot o(m,n-1)\,. \tag{E3.2}$$

Figure 4.3a graphically illustrates this equation when $m=n=2$. The circled point indicates the index for which the output of (E3.2) is to be computed. The X's indicate which inputs and previously computed outputs are needed for this computation. Notice that the output at $o(m,n)$ is always a function of the output immediately to the left, $o(m-1,n)$, or immediately under it $o(m,n-1)$. A consistent set of initial conditions is $o(m,n)=0$ for $m<0$. Any larger border regions of zeros for $o(m,n)$, e.g., the fourth quadrant, will not satisfy (E3.2) because $i_1(m,n)\neq 0$ for $m\geqq 0$. After a few attempts of computing $o(m,n)$ using (E3.2) the reader should realize that it is impossible. For instance, in order to compute $o(0,0)$, $o(0,-1)$ must be known and the calculation of $o(0,-1)$ requires $o(0,-2)$, etc. (see Fig. 4.3b). There exists no place to start the recursion without assuming a value for the output before it is computed. Therefore, for input $i_1(m,n)$ (E3.2) is not recursible; that is, there exist values of (m,n) where (E3.2) cannot compute the output; in fact, $o(m,n)$ cannot be computed using (E3.2) for any $m\geqq 0$.

There do exist examples where $o(m,n)$ can be computed consistently for some index values and not for others. For example, let $i_2(m,n)$ have support as shown in Fig. 4.3c so that $i_2(m,n)\neq 0$ for $(m,n)\in\{m\geqq 0 \text{ or } -m\geqq n\}$. If zero initial conditions on the set $\{(m,n): m<0 \text{ and } n<0 \text{ or } n\geqq 0 \text{ and } -m<n\}$ are assumed then any value of $o(m,n)$ for $(m,n)\in\{-m\geqq n \text{ and } m<0\}$ can be computed using (E3.2). For instance, $o(-1,3)$ can be computed if $o(-2,3)$ and $o(-1,2)$ are known. These later quantities can be solved if $o(-1,1)$ $o(-2,2)$ $o(-3,3)$ are known. Fortunately those three terms can be computed directly from (E3.2) as $i_2(-1,1)$, $i_2(-2,2)$, and $i_2(-3,3)$, respectively. Therefore, following arguments similar to this, one can show that any $o(m,n)$ for $(m,n)\in\{-m\geqq n \text{ and } m<0\}$ can be computed using (E3.2). Furthermore, the output values calculated from (E3.2) should be the same as these found

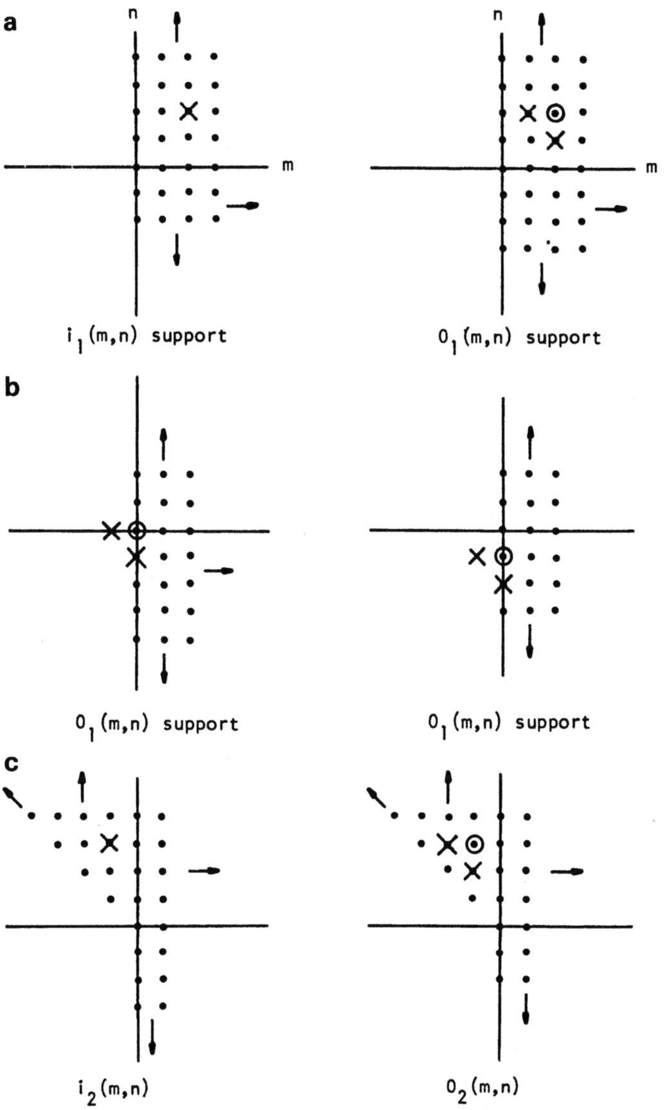

Fig. 4.3a–c. Examples depicting the effect of infinite input support on recursibility. (**a**) Computing $o(2,2)=i_1(2,2)+e\cdot o_1(m,n)+f\cdot o_1(2,1)$ where the circled support point indicates the value to be computed and the "X" points indicate known values needed in this computation; (**b**) where should the recursion start? (**c**) $O_2(m,n)$ can be computed recursively for $m<0$ but not for $m\geqq 0$

convolving the impulse response (E3.1) with $i_2(m,n)$. Unfortunately, (E3.2) cannot be used in this case if $m\geqq 0$ because the same problem as above arises. Therefore, (E3.1) is not recursible for $i_2(m,n)$ since there exist indices where $o(m,n)$ cannot be computed.

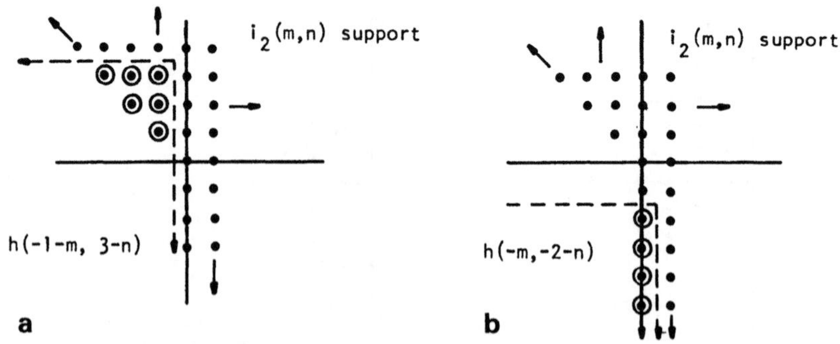

Fig. 4.4a, b. For $m<0$ convolution sum always has a finite number of terms while for $m \geqq 0$ it always has an infinite number of terms in this example. (**a**) Finite number of terms in convolution sum; (**b**) infinite number of terms in convolution sum

These examples give some insight on why recursive equations are not recursively computable for some inputs. Recall from (4.3) that $o(m,n) = h(m,n)*i(m,n)$ where $*$ denotes convolution and $h(m,n)$ is the impulse response of the filter. In order to compute $o(m,n)$ at $(m,n)=(m',n')$ by the convolution sum the impulse response is first inverted with respect to the m and n axes; then it is shifted by (m',n') and placed on top of $i(m,n)$; all overlying values of h and i are multiplied and summed giving $o(m',n')$. Figure 4.4a illustrates the convoluton of $i_2(m,n)$ with $h(m,n)$ of (E3.1) for $(m',n')=(-1,3)$. Notice that for any finite, $m<0$ this sum always has a finite number of nonzero terms in the sum. However, for any $m' \geqq 0$ this sum will have an infinite number of nonzero terms (see Fig. 4.4b). The convolution sum is well defined in either case although no closed form representation may exist for the latter case. However, since the recursive equation (4.9) is restricted to finite sums it cannot compute values of $o(m,n)$ where the convolution sum requires an infinite sum. Therefore, if the support of $h(m'-m, n'-n)$ (function of m and n) ever intersects the support of $i(m,n)$ at an infinite number of points then (4.9) is not recursible. In other words, a recursive filter with impulse response $h(m,n)$ cannot be used to filter $i(m,n)$ for all values of (m,n). It is not difficult to show that this condition can be equivalently stated in terms of β, the support set of $b(m,n)$ as follows: β^*, the smallest section of lattice points containing β must not intersect the support of $i(m-r, n-s)$ (function of r and s) at an infinite number of points for any (m,n). The following theorem summarizes the above discussions:

Theorem 1. Let $a(m,n)$ and $b(m,n)$ be two real finite extent arrays satisfying (a) $b(0,0)=1$ and (b) there exists an $(m,n) \neq (0,0)$ such that $b(m,n) \neq 0$. Then

$$o(m,n) = \sum_{(r,s)} a(r,s) i(m-r, n-s) - \sum_{(k,l) \in \beta - (0,0)} b(k,l) o(m-k, n-l)$$

is recursible if and only if

(1) $\beta = \{(m,n) : b(m,n) \neq 0\}$ is a subset of a lattice sector with vertex $(0,0)$ of angle less than π,

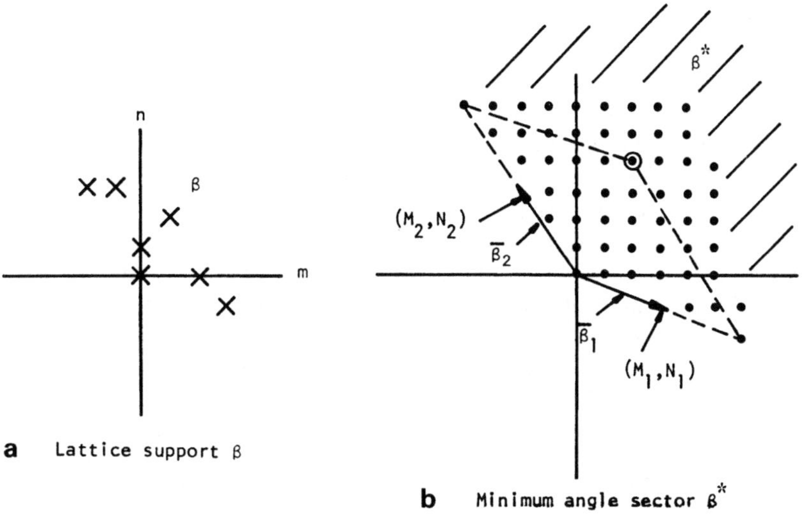

Fig. 4.5a, b. Typical β and β^* where $\beta^* = S[(3,-1),(-2,3)]$. (**a**) Lattice support β; (**b**) minimum angle sector β^*

(2) the support of $i(m-r, n-s)$ when considered as a function of (r,s) does not intersect β^* (which is the minimum angle lattice sector containing β) at an infinite number of points for all finite (m,n).

If $b(m,n)$ satisfies the conditions in the above theorem, that is (a), (b), and (1), then it will be called a *recursive filter array*. Associated with a recursive filter array is a lattice sector called β^* which consists of all lattice points in the minimum angle sector containing β. β^* can be uniquely defined by two vectors $\bar{\beta}_1 = (M_1, N_1) \in Z^2$ and $\bar{\beta}_2 = (M_2, N_2) \in Z^2$ as $\beta^* = S[(M_1, N_1), (M_2, N_2)]$ $= \{(m,n) \in Z^2 : (m,n) = r_1\bar{\beta}_1 + r_2\bar{\beta}_2$, with r_1 and r_2 nonnegative real numbers$\}$ where Z is the set of integers and M_1 and N_1 along with M_2 and N_2 are mutually prime integers. If $M_1N_2 - N_1M_2 \neq 0$ then $\bar{\beta}_1$ and $\bar{\beta}_2$ are not colinear vectors and hence β is composed of noncolinear points. However, if $M_1N_2 - N_1M_2 = 0$ then $\bar{\beta}_1$ and $\bar{\beta}_2$ are colinear (in fact, equal) and β is composed entirely of colinear points. Figure 4.5a illustrates β the support of a typical recursive filter array and Fig. 4.5b shows $\beta * = S[(3,-1),(-2,3)]$ the minimum angle sector containing β.

Since the $a(m,n)$ array does not affect recursibility, it will be assigned the value $\delta(m,n)$ for the remainder of the discussion. This simplification will allow us to avoid some rather unpleasant problems due to nonessential singularities of the second kind which may arise when discussing stability [4.9, 10]. Equations (4.9, 10) now become

$$o(m,n) = i(m,n) - \sum_{(r,s)\in\beta-(0,0)} b(r,s)o(m-r, n-s) , \tag{4.11}$$

$$h(m,n) = \delta(m,n) - \sum_{(r,s)\in\beta-(0,0)} b(r,s)h(m-r, n-s) . \tag{4.12}$$

So far it has been stated that under appropriate conditions (4.9–12) are recursible; however, no mention has been made about how to recurse. For simplicity, we shall consider only (4.12). First, a close study of this equation will reveal that $h(m,n)=0$ for $(m,n)\in Z^2-\beta^*$ is a consistent set of initial conditions where Z^2 is the set of all integer ordered pairs. Next, a sufficient condition for calculating $h(m,n)$ with $(m,n)\in\beta^*$ is to know all $(h(m,n)$ in the parallelogram formed by connecting this point to the rays formed from $\bar{\beta}_1$ and $\bar{\beta}_2$ [see Fig. 4.5b]. Here, points and lines are considered to be degenerate parallelograms. Therefore, starting with $h(0,0)=1$ we can build up any parallelogram by a number of smaller parallelograms, and hence, we can find any $h(m,n)$ in this manner. The "parallelogram rule" presents one approach of computing $h(m,n)$ for any $(m,n)\in\beta^*$. This rule can be extended to apply to (4.9–11). From the definition of β^* it is not difficult to show that there exist positive integers r_1 and r_2 such that $h(r_1M_1,r_1N_1)\neq 0$ and $h(r_2M_2,r_2N_2)\neq 0$, and thus, β^* is the minimum angle sector containing the infinite extent impulse response $h(m,n)$ [from (4.12)]. In general, (4.9, 11) usually implement LSI systems with infinite extent impulse responses which have supports which are subsets of shifted sectors of angle less than π [4.1].

There exists another method of finding the recursive solution of (4.12) which will prove useful in the forthcoming stability discussion. Define $B(w,z)$ for a recursive filter array $b(m,n)$ as

$$B(w,z)=\sum_{(r,s)\in\beta} b(r,s)w^r z^s\,, \tag{4.13}$$

where r and s can be negative as well as positive integers. Since $b(0,0)=1$, (4.13) can be written as $B(w,z)=1-C(w,z)$, where $C(w,z)$ is a polynomial with no constant term. A formal expansion for $1/B(w,z)$ can be obtained by

$$1/B(w,z)=\frac{1}{1-C(w,z)}=\sum_{n=0}^{\infty}[C(w,z)]^n=\sum_{(m,n)\in\beta^*} h(m,n)w^m z^n\,. \tag{4.14}$$

This is strictly a formal power series and no claim of any type of convergence is given. Since $C(w,z)$ has no constant term and its coefficient array is confined to a sector of angle less than π this implies that for any $n>0,[C(w,z)]^n$ can only contribute to terms $h(m,n)$ supported in β^*. Furthermore, given any $(m',n')\in\beta^*$ there exists an $n_1>0$ such that $\sum_{n=n_1}^{\infty}[C(w,z)]^n$ does not contribute to the $h(m',n')w^{m'}z^{n'}$ term of $\sum h(m,n)w^m z^n$. In other words, any $h(m,n)$ can be found in a finite number of calculations. This will be discussed further after example 4. We claim that the sequence $h(m,n)$ obtained via (4.14) is equal to that obtained from (4.12). The following theorem establishes this.

Theorem 2. Assume $b(m,n)$ is a recursive filter array, $h_1(m,n)$ is the solution of (4.12) [remember, this is the solution acquired by appropriately iterating (4.12)] and that $h_2(m,n)$ is the sequence obtained from (4.14), then $h_1(m,n)=h_2(m,n)$.

Proof. First, both sequences were constructed so that they are solutions to the following difference equation:

$$\sum_{(r,s)\in\beta} b(r,s)h(m-r,n-s)=\delta(m,n)\ .$$

Therefore, $f(m,n)=h_1(m,n)-h_2(m,n)$ must be a particular solution of the homogeneous equation

$$\sum_{(r,s)\in\beta} \mathrm{b(r,s)\,f(m-r,n-s)=0}\ . \tag{4.15}$$

Secondly, since h_1 and h_2 have support in β^*, $f(m,n)$ must be supported there also. Therefore, (4.15) means that the convolution of the two arrays $b(r,s)$ and $f(r,s)$, both with supports in β^*, is always zero. Recalling that $b(0,0)=1$ and that β^* is a lattice sector at angle less than π, it follows that $f(0,0)=0$, because β intersects the support of $f(-r,-s)$ only at $(0,0)$. Next, (4.15) can be used to show that $f(M_1,N_1)$ and $f(M_2,N_2)$ must also be zero. Proceeding successively in this fashion it can be shown that $f(m,n)=0$ for all $(m,n)\in\beta^*$. Therefore, $h_1(m,n)=h_2(m,n)$. QED

A third method of calculating $h(m,n)$ of (4.12) is to employ a division algorithm. This was proposed by *Dudgeon* [4.1]. First, form $B(w,z)$ $=\sum_{(m,n)\in\beta} b(m,n)w^m z^n$. Next, formally divide $B(w,z)$ into 1 [in the more general case divide $A(w,z)$ by $B(w,z)$] in such a fashion that the constant term is used to eliminate terms in the remainder one after another. When carried out properly the resulting quotient will be the series expansion of $H(w,z)$. Hence by (4.5), $h(m,n)$ is the coefficient of $w^m z^n$ term. One difficulty in performing this algorithm is determining whether the coefficients in the quotient are finalized. In other words, in what order should the monomials in the remainder be eliminated to ensure that it never appears in the remainder or quotient again? It can be shown that many such orders exist and the parallelogram rule specifies some of them. However, this is only true if β^* has angle less than π. If this is not the case then some of the coefficients of terms in the quotient cannot be finalized in a finite number of steps.

Example 4 (E4)

Below, it will be shown by means of a simple example how each of these algorithms can calculate $h(m,n)$ of (4.12). Let $b(m,n)$ be defined as follows: $b(0,0)=1$, $b(1,0)=-1$, $b(2,-1)=-1$ and $b(-1,1)=-1$. Notice that $b(m,n)$ is a recursive filter array and so by (4.12)

$$h(m,n)=\delta(m,n)+h(m-1,n)+h(m+1,n-1)+h(m-2,n+1)\ . \tag{E4.1}$$

On the other hand, this recursive filter can be represented by $B(w,z)=1-w-w^{-1}z-w^2z^{-1}$ which is equal to $1-C(w,z)$ where $C(w,z)=w+w^{-1}z+w^2z^{-1}$.

Method one: Recursive Equation

The above recursive equation along with the assumption of zero initial conditions is utilized to solve for $h(m,n)$.

$$\begin{aligned}
h(0,0) &= 1+h(-1,0)+h(1,-1)+h(-2,1)=1\\
h(-1,1) &= h(-1,1)+h(0,0)+h(-3,2)=1\\
h(-2,2) &= h(-3,2)+h(-1,1)+h(-4,3)=1\\
h(-3,3) &= h(-4,3)+h(-2,2)+h(-5,4)=1\\
h(-m,m) &= \qquad\qquad\qquad\qquad =1 \quad (m\geqq 0)\\
h(2,-1) &= h(1,-1)+h(3,-2)+h(0,0)=1\\
h(1,0) &= h(0,0)+h(2,-1)+h(-1,1)=3\\
h(0,1) &= h(-1,1)+h(1,0)+h(-2,2)=5\\
h(-1,2) &= h(-2,2)+h(0,1)+h(-3,3)=7\\
h(1-m,m) &= 1+2(m+1) \qquad (m\geqq -1)\\
h(2-m,m) &= 1+2(m+2)(m+3) \qquad (m\geqq -2)
\end{aligned}$$

etc.

Method two: Division Algorithm

$$\begin{array}{r|l}
 & 1+w^{-1}z+w^{-2}z^2+w^{-3}z^3+w^2z^{-1}+3w+5z+7w^{-1}z^2\\
\hline
1-w^{-1}z-w-w^2z^{-1} & 1\\
 & \underline{1-w^{-1}z-w-w^2z^{-1}}\\
 & w^{-1}z+w+w^2z^{-1}\;.
\end{array}$$

etc.

Method three: Geometric Expansion

$$\frac{1}{B(w,z)}=\sum_{n=0}^{\infty}[C(w,z)]^n\,.$$

By considering only four terms of this series we obtain

$$\begin{aligned}
&1+C(w,z)+[C(w,z)]^2+[C(w,z)]^3\\
&=1+(w^{-1}z+w^2z^{-1}+w)+(w^{-2}z^2+w^2+w^4z^{-2}+2z+2w+2w^3z^{-1})\\
&\quad+(w^{-3}z^{-3}+w^3+w^6z^{-3}+6w^2+3w^{-1}z^2+3wz+3z+3w^3z^{-1}\\
&\quad+3w^4z^{-1}+3w^5z^{-2})\,.
\end{aligned}$$

A close look will reveal that this expansion only contains seven finalized monomials given by

$$1+w^{-1}z+w^{-2}z^2+w^{-3}z^3+w^2z^{-1}+3w+5z,$$

where the coefficient of $w^m z^n$ is equal to $h(m,n)$. In general, since β is contained in a lattice sector of angle less than π with $C(w,z)$ having no constant term and since monomial multiplication is equivalent to vector addition then these facts imply that each new terms in the geometric expansion finalizes more coefficients. This vector addition approach to recursive filtering leads to an interesting interpretation which is presented in the following example.

Example 5 (E5)

In the above discussion on the geometric expansion method for calculating the impulse response we alluded to a vector interpretation of recursive filtering. This interpretation was motivated by the fact that polynomial multiplication in two variables (n variables) can be viewed as a special kind of weighted vector addition. For example, let $B_1(w,z)=w+wz$ and $B_2(w,z)=w^{-1}+w^{-1}z+wz$ and calculate

$$C(w,z)=B_1(w,z)\cdot B_2(w,z)=1+2z+w^2z+z^2+w^2z^2. \qquad \text{(E5.1)}$$

Figure 4.6 graphically illustrates that this product can be formed by first obtaining all vector sums of individual monomials interpreted as vectors from B_1 and B_2. Next, assign a coefficient to the resulting vector (or lattice point) equal to the sum of all the products of vector pairs terminating at that point. This procedure is also valid for any kind of polynomial multiplication. Furthermore, it can be applied to recursive filters because (2.14) relates the geometric series $\sum_{n=0}^{\infty} [C(w,z)]^n$ to $H(w,z)$. In this case each monomials of $C(w,z)$ represents a weighted vector. Moreover, since none of these vectors differ by an angle greater than or equal to π, then the impulse response at a point (r,s) is

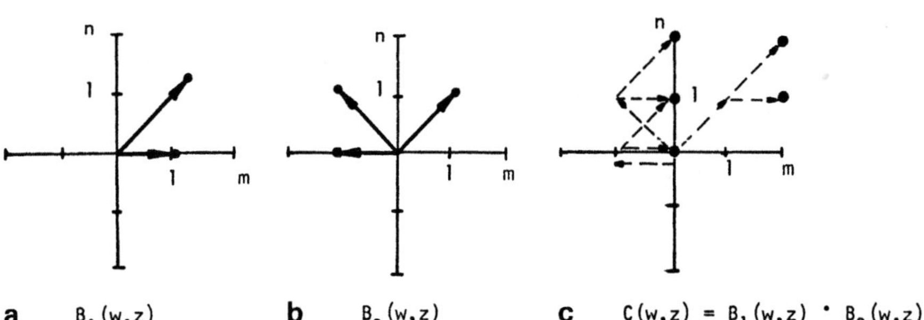

Fig. 4.6a–c. Vector addition approach for polynomial multiplication. **(a)** $B_1(w,z)$; **(b)** $B_2(w,z)$; **(c)** $C(w,z)=B_1(w,z)\cdot B_2(w,z)$

always formed by a finite number of vector paths. Therefore, for any given (r, s) there exists a finite largest exponent of $C(w, z)$ in (2.14) which contributes to the vector (r, s), that is, to the monomial $w^r z^s$. Hence, any impulse response value can be finalized by using the geometric expansion approach. However, if the restriction that β lie in a sector of angle less than π is removed, then the finality of coefficients conditions is no longer valid. This, in turn, implies that recursive filtering cannot be performed.

This vector addition interpretation to recursive filtering can give practical insight into the application of simple filters. For example, in many instances it may be obvious that no vector path of a given filter can produce a certain lattice point. If this is the case then the impulse response value at that index point is zero. For instance, let $B(w, z) = l + w^2 + z^2 + w^2 z^2$. Here, all impulse response values with at least one odd index will be zero since the vectors representing w^2, z^2, and $w^2 z^2$ can never add up to such a vector. More complicated examples can be easily constructed.

4.4. Stability Theorems

In most applications only bounded-input bounded-output (BIBO) stable filters are of interest. If a filter is unstable, any noise, including round-off errors in computation, will propagate through the output and be amplified. As stated earlier, a necessary and sufficient condition for BIBO stability of a LSI system is that $\sum_{(m,n)} |h(m, n)| < \infty$. Therefore, the recursive filter represented by the recursive filter array $b(m, n)$ is stable if and only if the impulse response obtained from (4.12) is absolutely summable, that is, $\sum_{(m,n)\in\beta^*} |h(m, n)| < \infty$. This impulse response can be rotated by a multiple of 90°, transposed or flipped and the resulting sequences are still absolutely summable. Equation (4.14) can be employed to identify the resulting recursive filter array. These facts can be utilized to readily prove the following theorem which is a generalization of results given in [4.11].

Theorem 3. Let $b(m, n)$ be a stable recursive filter array, then all the following recursive filter arrays are also stable

a) $b(m, n)$ identity
b) $b(n, -m)$ clockwise 90° rotation
c) $b(-m, -n)$ clockwise 180° rotation
d) $b(-n, m)$ clockwise 270° rotation
e) $b(-m, n)$ n-axis mirror image
f) $b(-n, -m)$ transpose with respect to $m = -n$ diagonal
g) $b(m, -n)$ m-axis mirror image
h) $b(n, m)$ transpose with respect to $m = n$ diagonal.

This theorem is a special case of a more general mapping theorem which will be developed below.

Consider two sectors $S_1[(M_1, N_1), (M_2, N_2)]$ and $S_2[P_1, Q_1), (P_2, Q_2)]$ with $D = M_1N_2 - M_2N_1 \neq 0$ and $E = P_1Q_2 - P_2Q_1 \neq 0$. There exists many linear mappings of the form

$$\begin{aligned} m &= k_1m' + k_2n' \\ n &= k_3m' + k_4n' \end{aligned} \quad (k_i \in Z) \tag{4.16}$$

that map S_1 one-to-one (not necessarily onto) into S_2. For example, one mapping is as follows:

$$\begin{aligned} k_1 &= (N_2P_1 - N_1P_2)\,\mathrm{sgn}(D) \\ k_2 &= (M_1P_2 - M_2P_1)\,\mathrm{sgn}(D) \\ k_3 &= (N_2Q_1 - N_1Q_2)\,\mathrm{sgn}(D) \\ k_4 &= (M_1Q_2 - M_2Q_1)\,\mathrm{sgn}(D) \end{aligned}$$

with this mapping S_2 is the smallest sector containing the mapped points. If S_2 is the first quadrant in the lattice plane, then (4.16) can be used to derive the following mapping of S_1 into S_2 by setting $P_1 = 1$, $Q_1 = 0$, $P_2 = 0$, and $Q_2 = 1$:

$$\begin{aligned} k_1 &= \mathrm{sign}(D)N_2 \\ k_2 &= -\mathrm{sgn}(D)M_2 \\ k_3 &= -\mathrm{sgn}(D)N_1 \\ k_4 &= \mathrm{sgn}(D)M_1\,. \end{aligned} \tag{4.17}$$

These sector transformations are similar to mappings given recently in [4.12]. In [4.12], *Dudgeon* employed a transformation similar to 4.16) to compute the cepstra of non-quadrant arrays. The above ideas will be utilized to prove the following important theorem:

Theorem 4. Let $b(m,n)$ be a recursive filter array with support β. Then $b(m,n)$ is stable if and only if for any $k_i \in Z$ with $K = k_1k_4 - k_2k_3 \neq 0$ the recursive filter array

$$g(m,n) = g(k_1m' + k_2n', k_3m' + k_4n') = b(m',n') \tag{4.18}$$

with $(m',n') \in \beta$ is stable

Proof. Since $K \neq 0$ then $g(m,n)$ is a recursive filter array with support

$$\gamma = \{(m,n) \in Z^2 : m = k_1m' + k_2n',\, n = k_3m' + k_4n',\, (m',n') \in \beta\}\,.$$

The requirement that $K \neq 0$ also implies that the mapping (4.16) of $b(m,n)$ to $g(m,n)$ is one-to-one and onto. Let $h_1(m,n)$ and $h_2(m,n)$ be the impulse responses of filters $b(m,n)$ and $g(m,n)$ respectively, with respective supports of β' and γ'. (Note that β' and γ' are subsets of β^* and γ^*, respectively.) It will first be shown that $h_2(k_1m+k_2n,\ k_3m+k_4n)=h_1(m,n)$ for $(m,n)\in\beta'$. Define $G(w,z) = \sum_{(m,n)\in\gamma} g(m,n)w^m z^n$. Therefore,

$$G(w,z)=\sum_{(m,n)\in\gamma} g(m,n)w^m z^n = \sum_{(m,n)\in\beta} g(k_1m+k_2n, k_3m+k_4m)w^{k_1m+k_2n}z^{k_3m+k_4n}$$

$$=\sum_{(m,n)\in\beta} b(m,n)\,(w^{k_1}z^{k_3})^m(w^{k_2}z^{k_4})^n$$

$$=B(w^{k_1}z^{k_3}, w^{k_2}z^{k_4})\,.$$

By (4.14) and Theorem 2 it follows that

$$1/G(w,z)=\sum_{(m,n)\in\gamma'} h_2(m,n)w^m z^n$$

$$=1/B(w^{k_1}z^{k_3}, w^{k_2}z^{k_4})=\frac{1}{1-C(w^{k_1}z^{k_3}, w^{k_2}z^{k_4})}$$

$$=\sum_{(m',n')\in\beta'} h_1(m,n)\,(w^{k_1}z^{k_3})^m(w^{k_2}z^{k_4})^n$$

$$=\sum_{(m,n)\in\beta'} h_1(m,n)w^{k_1m+k_2n}z^{k_3m+k_4n}\,.$$

Since $K\neq 0$, for each $(m,n)\in\beta'$, the term $h_1(m,n)w^{k_1m+k_2n}z^{k_3m+k_4n}$ of the last series only occurs once. Hence, it follows that $h_1(m,n)=h_2(k_1m+k_2n,\ k_3m+k_4n)$ for all $(m,n)\in\beta'$. This implies that impulse responses $h_1(m,n)$ and $h_2(m,n)$ are related by a one-to-one and onto mapping, so if one is absolutely summable the other will be and visa versa. QED

This theorem is important because it allows virtually any type of recursive filter to be mapped into virtually any other type while at the same time preserving stability. In particular, if $D=M_1N_2-M_2N_1\neq 0$ then (4.17) can be used to map a general recursive filter to a first-quadrant quarter-plane filter with $K=D\neq 0$. Hence, the stability of a non-first-quadrant array can be determined by checking the stability of the first-quadrant filter obtained by mapping $b(m,n)$ into the first quadrant. Equation (4.17) offers one such mapping if β has at least three noncolinear points. However, if β is composed of entirely colinear points on the line defined by (M_1, N_1) (mutually prime integers) then $b(m,n)$ can be mapped into a one-dimensional filter sequence $c(m)$ by $c(m)=b(mM_1, mN_1)$. It's not difficult to show that the resulting one-dimensional filter is stable if and only if the original filter is stable.

Example 6 (E6)

As an example, consider the recursive filter array $b(m,n)$ defined as

$$B(w,z)=0.5w^{-1}z+1+0.85w+0.1wz+0.5w^2z^{-1}. \tag{E6.1}$$

Note that $\beta^*=S[(2,-1),(-1,1)]$. $b(m,n)$ can be mapped into a first-quadrant filter, say $g(m,n)$, using (4.16, 17). This yields $g(m,n)=g(m'+n',m'+2n')=b(m',n')$ where $(m',n')\in\beta$. Therefore,

$$G(w,z)=1+0.5w+0.5z+0.85wz+0.1w^2z^3. \tag{E6.2}$$

By Theorem 4, the stability of $b(m,n)$ can be determined by checking the stability of $g(m,n)$, a first-quadrant filter.

There exist a number of theorems for determining the stability of first-quadrant quarter-plane filters. All of them relate stability to the zero set of the equation $B(w,z)=0$. The following theorem summarizes these results.

Theorem 5. Let $b(m,n)$ be a first-quadrant quarter-plane recursive filter array and define $B(w,z)=\sum_{(m,n)\in\beta} b(m,n)w^mz^n$. The two-dimensional recursive filter $b(m,n)$ is stable if and only if

1) $B(w,z)\neq 0\,|w|\leqq 1\,|z|\leqq 1$

if and only if (iff)

2) (a) $B(w,z)\neq 0\,|w|\leqq 1\,|z|=1$
   (b) $B(a,z)\neq 0\,|z|\leqq 1$ for some a, $|a|\leqq 1$

iff

3) (a) $B(w,z)\neq 0\,|w|=1\,|z|\leqq 1$
   (b) $B(w,b)\neq 0\,|w|\leqq 1$ for some b, $|b|\leqq 1$

iff

4) (a) $B(w,z)\neq 0$ on $T^2=\{(w,z):|w|=1=|z|\}$
   (b) $B(a,z)\neq 0\,|z|\leqq 1$ for some a, $|a|\leqq 1$
   (c) $B(w,b)\neq 0\,|w|\leqq 1$ for some b, $|b|=1$

iff

5) (a) $B(w,z)\neq 0$ on T^2
   (b) $B(z,z)\neq 0\,|z|\leqq 1$

iff

6) (a) $B(w,z)\neq 0$ on T^2
   (b) $B(z^{l_1},z^{l_2})\neq 0\,|z|\leqq 1$

for some $l_1, l_2\in Z^+$ (where Z^+ denotes the positive integers).

Proof. 1) See [4.3–5, 13].

2)–3) *Huang* [4.3] showed this for $a=0$ ($b=0$) and it can easily be shown for $|a|\leqq 1$ or $|b|\leqq 1$ by modifying the proof in [4.14]. Also see [4.15].

4) See *DeCarlo* [4.6], *Strintzis* [4.9, 15], following discussion, and [4.16].

5) See [4.6].

6) This is a generalization of a theorem [4.6] by *DeCarlo* et al. QED

With this theorem as an introduction it seems appropriate to review the history of stability theorems which relate stability to the zero set of $B(w,z)$, *Shanks* et al. [4.4] were apparently the first researchers to seriously study both the stability and synthesis of two-dimensional recursive filters. In their first paper [4.4] which considered only first-quadrant filters they presented a theorem equivalent to Theorem 5.1. In a later paper [4.5] a rigorous mathematical proof for this based on communative Banach algebra theory and a Tauberian theorem was presented. Surprisingly, the theorem allows $b(m,n)$ to have an infinite number of nonzero entries, however, this extension has little practical value. Furthermore, they proved theorems relating the zero regions of $B(w,z)$ to the stability of half-plane and full-plane filters. However, these more general filters were not necessarily recursible.

At about the same time *Huang* [4.3] proved theorems equivalent to Theorems 5.2 and 5.3. Theorem 5.3 greatly simplified the stability conditions of Shank's Theorem. Virtually all stability tests which followed tried to check the conditions of this theorem. Apparently, in the proof of this theorem several minor errors were made. *Davis* [4.17] and *Goodman* [4.14] discovered these errors and independently reproved the theorem. Later, *DeCarlo* et al. [4.18, 19] and *Strintzis* [4.15] presented similar and interesting proofs of this theorem. *DeCarlo* used homotopy theory, covering theory, and an extension of Nyquist theory to validate the theorem. He later [4.6] extended these ideas to prove Theorems 5.4 and 5.5. *Strintzis* [4.15] gave very simple proofs of Theorems 5.2, 5.3, and 5.4 using the Cauchy principal value formula which is

$$N(z_1)=\frac{1}{2\pi j}\int_{|z_2|=1}\frac{\partial B(z_1,z_2)}{\partial z_2}[B(z_1,z_2)]^{-1}dz_2 \tag{4.19a}$$

or

$$M(z_2)=\frac{1}{2\pi j}\int_{|z_1|=1}\frac{\partial B(z_1,z_2)}{\partial z_1}[B(z_1,z_2)]^{-1}dz_1\,, \tag{4.19b}$$

where $N(z_1)$ for a fixed value of z_1 is the number of zeros in z_2 of $B(z_1,z_2)$ that lie in the region $|z_2|\leqq 1$ and for z_2 fixed, $M(z_2)$ is the number of zeros in z_1 of $B(z_1,z_2)$ that lie in the region $|z_1|\leqq 1$. Furthermore, he used this approach to derive stability conditions for nonsymmetric half-plane filters. In our notation these are filters where the recursive filter array is confined to one of the eight nonsymmetric half planes (see Fig. 4.1). Nonsymmetric half-plane filters were first introduced by *Ekstrom* and *Woods* [4.8] and *Dudgeon* [4.1, 2]. *Ekstrom* and

Woods presented the first stability theorem for these types of filters. Each of the eight nonsymmetric filters had different stability conditions. Later, we will show that all these conditions can be simplified down to a few for any type of recursive digital filter including those more general than nonsymmetric half-plane filters. However, before considering these general theorems several interesting properties of first-quadrant two-dimensional recursive filters will be discussed.

4.5 An Interesting Property of Multidimensional Filters

A one-dimensional recursive filter can be represented by

$$H(z)=\frac{A(z)}{B(z)}, \tag{4.20}$$

where $A(z)=\sum_{m=0}^{M} a(m)z^m$ and $B(z)=\sum_{n=0}^{N} b(n)z^n$ or by the recursive equation

$$o(m)=\sum_{n=0}^{M} a(n)i(m-n)-\sum_{n=1}^{N} b(n)o(m-n), \tag{4.21}$$

where $b(0)=1$. The impulse response can be found by setting $i(m)=\delta(m)$ giving

$$h(m)=a(m)-\sum_{n=1}^{N} b(n)h(m-n). \tag{4.22}$$

This equation can be used to calculate any $h(m)$ for $m\geqq 0$ if zero initial conditions are assumed. If is well known that if $A(z)$ and $B(z)$ have no common zeros then the recursive filter is BIBO stable if and only if $B(z)\neq 0\,|z|\leqq 1$; that is $\sum_{m=0}^{\infty} |h(m)|<\infty (h\in l^1)$ if and only if (4.20) has no poles inside or on the unit circle. Therefore, for a one-dimensional filter the $a(m)$ array does not affect stability as long as $A(z)$ and $B(z)$ are relatively prime.

However, in two dimensions this is not the case. *Goodman* [4.10] has proven recently that if $A(w,z)\neq 1$ then Theorem 5 is incorrect. In particular, it is not a necessary condition that $B(w,z)\neq 0$ on $\{(w,z):|w|\leqq 1, |z|\leqq 1\}=\bar{U}^2$ although it is necessary that $B(w,z)\neq 0$ on $\{|w|\leqq 1, |z|\leqq 1\}-\{|w|=1=|z|\}$. Examples can be constructed where $A(w,z)$ and $B(w,z)$ are relatively prime and there exists a $(w_0,z_0)\in T^2=\{(w,z):|w|=1=|z|\}$ such that $A(w_0,z_0)=B(w_0,z_0)=0$ so that the filter is stable. In [4.10]

$$G_1(w,z)=\frac{(1-w)^8(1-z)^8}{1-\frac{1}{2}w-\frac{1}{2}z}=\frac{A_1(w,z)}{B_1(w,z)}$$

was found to be stable. Notice that $A_1(w,z)$ and $B_1(w,z)$ have no common factors and that $A_1(1,1)=B_1(1,1)=0$. In the complex variables literature the

(1, 1) point in this example is referred to as a nonessential singularity of the second kind. Another example with a similar singularity is

$$G_2(w,z)=\frac{(1-w)(1-z)}{1-\frac{1}{2}w-\frac{1}{2}z}\,.$$

In [4.10] it was shown that this filter is unstable. Therefore, given that $B(w,z)\neq 0$ on $\bar{U}^2-T$ and there exists a nonessential singularity of the second kind on T^2 then the filter may or may not be stable. At present, no easy procedure has been found to determine whether nonessential singularities of the second kind exist on T^2 for a given filter. Moreover, it is also very difficult to determine stability given that a nonessential singularity of the second kind occurs on T^2.

It is not difficult to show that $A(w,z)$ may affect the stability of general two-dimensional recursive filters. This is why that in the statements of all the stability theorems it was assumed that $A(w,z)=1$. However, it appears that in practice, recursive filter design algorithms will virtually never produce non-essential singularities on T^2 at the same time satisfying $B(w,z)\neq 0$ on $\bar{U}^2-T^2$.

4.6 General Stability Theorems

In this section several theorems which directly relate the stability of a general recursive filter, represented by a recursive filter array $b(m,n)$, to the zero distribution of $B(w,z)$ will be derived. It seems that in the past only *Ekstrom* and *Woods* [4.8] and *Strintzis* [4.15] have considered this extension and they worked with the stability of nonsymmetric half-plane filters which are a subclass of the considered filters.

Theorem 6. Let $b(m,n)$ be a recursive filter array with support β and associated minimum angle sector $\beta^*=S[(M_1,N_1),(M_2,N_2)]$. Then $b(m,n)$ is stable if and only if

Case 1: $D=M_1N_2-M_2N_1\neq 0$

(a) $B(w,z)\neq 0$ on T^2
(b) $B(\lambda^{P_1},\lambda^{P_2})\neq 0$ for $|\lambda|\leqq 1$ for some P_1 and P_2
where $P_1=\mathrm{sgn}(D)N_2l_1-\mathrm{sgn}(D)N_1l_2$
$P_2=-\mathrm{sgn}(D)M_2l_1+\mathrm{sgn}(D)M_1l_2$
and $l_i\in Z^+$ (where Z^+ denotes the positive integers).

Case 2: $D=0$

(a) for some $l_1\in Z^+$

$$C(\lambda)=\sum_{m=0} c(m)\lambda^{l_1 m}\neq 0 \quad \text{for} \quad |\lambda|\leqq 1$$

where $c(m)=b(mM_1,mN_2)$.

Proof. Case 1: $D \neq 0$. By Theorem 4, the first-quadrant filter $g(m,n) = g(k_1 m' + k_2 n',\ k_3 m' + k_4 n') = b(m',\ n')$ where k_i, $i=1,\ldots,4$ are given in (4.17) is stable iff $b(m,n)$ is stable. By part 6) of Theorem 5 $g(m,n)$ is stable iff (i) $G(w,z) \neq 0$ on T^2, (ii) $G(\lambda^{l_1}, \lambda^{l_2}) \neq 0$ for $|\lambda| \leq 1$ for some $l_1, l_2 \in Z^+$. Equation (4.18) and Theorem 4 imply that $G(w,z) = B(w^{k_1} z^{k_3}, w^{k_2} z^{k_4})$, and therefore, since $K = k_1 \cdot k_4 - k_2 \cdot k_3 = D \neq 0$ with $k_i \in Z$ then $G(w,z) \neq 0$ on T^2 iff $B(w,z) \neq 0$ on T^2. Moreover, part (b) follows directly from (ii) after the appropriate change of variables.

Case 2: $D=0$. This follows directly from the discussion after Theorem 4 and results on the stability of one-dimensional recursive filters [4.20]. QED

Example 7 (E7)

Consider the recursive filter given in Example 6 where

$$B(w,z) = 0.5w^{-1}z + 1 + 0.85w + 0.1wz + 0.5w^2 z^{-1}.$$

According to Theorem 6 the filter $b(m,n)$ is stable if and only if (a) $B(w,z) \neq 0$ on T^2 and (b) $B(\lambda^{P_1}, \lambda^{P_2}) \neq 0$ for $|\lambda| \leq 1$ for some P_1 and P_2 satisfying the given conditions. It is easy to show that $\beta^* = S[(2,-1),\ (-1,1)]$ so $D = M_1 N_2 - M_2 N_1 = 1 \neq 0$. Therefore, letting $l_1 = 1 = l_2$ condition (b) becomes

$$B(\lambda^2, \lambda^3) = 1 + \lambda + 0.85\lambda^2 + 0.1\lambda^5 \neq 0 \tag{E7.1}$$

on $|\lambda| \leq 1$. A number of different algorithms can be employed to show that (E7.1) satisfies condition (b). Testing condition (a) is much more difficult and will be considered later.

Before starting the next theorem a review of some facts concerning one-dimensional sequences will help clarify our terminology. Assume $d(n)$ is a finite sequence of real numbers, $d(M), \ldots, d(N)$ with $M \leq N$, and $d(M) \neq 0$ and $d(N) \neq 0$. A polynomial (possibly in both z and z^{-1}) $D(z) = \sum_{n=M}^{N} d(n) z^n$ can be formed which has P_1 roots counting multiple roots and roots at infinity where

$$P_1 = |\min(0,M)| + \max(0,N). \tag{4.23}$$

Let $M^1 = |\min(0,M)|$ and $N^1 = \max(0,N)$ so $P_1 = M^1 + N^1$. If $D(z)$ has no roots on the unit circle then we can define on $[-\pi, \pi]$ a continuous, odd phase function associated with the Fourier transform of the sequence. This phase function contains no linear phase component if and only if $D(z)$ has exactly N^1 zeros outside the unit cirle and M^1 zeros inside. This follows from the argument principle which can be stated as follows:

Argument Principle: [4.21] Let C be a simple contour described in positive sense and let f be a function which is analytic inside and on C, except possible for poles interior to C. Also let f have no zeros on C. Then

$$\frac{1}{2\pi}\Delta_c \arg f(z) = N - P, \qquad (4.24)$$

where N and P are the number of zeros and the number of poles of f, counting multiplicities, interior to C.

In particular, if C is the unit circle and if $D(z)$ has no zeros on C then there exist M^1 poles inside the unit circle (in fact, M^1 poles at zero). By the Argument Principle the total change of phase (arg) of $D(z)$ on C is

$$\Delta_c \arg D(z) = 2\pi(N - M^1),$$

where N is the number of zeros of $D(z)$ inside the unit circle. Therefore, the total change of phase of $D(z)$ is zero over C if and only if $N = M^1$, that is, $D(z)$ has exactly N^1 zeros outside the unit circle and M^1 zeros inside. Thus any time it is stated that a sequence or a polynomial has no linear phase component it means that it satisfies the above conditions.

The fact that the phase of the Fourier transform has no linear component can be interpreted with the aid of Nyquist theory. If the unit circle is defined to be the Nyquist contour the above phase condition means that the Nyquist plot (in this case, $D(e^{jw})$, $w=0$ to $w=2\pi$) does not encircle nor pass through the origin. In other words, $\mathrm{Ind}(D(e^{j2\pi s}))=0$ where Ind indicates the number of encirclements of the origin. Therefore, both interpretations are equivalent.

Theorem 7. $b(m,n)$ is a stable recursive filter array if and only if

(a) $B(w,z) \neq 0$ on T^2;
(b) $B(\lambda, 1)$ and $B(1, \lambda)$ have no linear phase terms
[i.e., $\mathrm{Ind}(B(\lambda,1)) = 0 = \mathrm{Ind}(B(1,\lambda))$].

Proof. Sufficiency: If part (a) holds, then it can be shown [4.22] that

$$\mathrm{Ind}[B(\lambda^{l_1}, \lambda^{l_2})] = l_1 \cdot \mathrm{Ind}[B(\lambda,1)] + l_2 \cdot \mathrm{Ind}[B(1,\lambda)] \qquad (4.25)$$

for any $l_1, l_2 \in Z$. Condition (b) implies that $\mathrm{Ind}[B(\lambda^{l_1}, \lambda^{l_2})] = 0$ for any $l_1, l_2 \in Z$, and, in particular, for $l_1 = P_1$ and $l_2 = P_2$ of Theorem 6. Therefore, since $B(\lambda^{P_1}, \lambda^{P_2})$ is a polynomial in positive powers of λ, this implies it has no zeros in $|\lambda| \leqq 1$. Hence, conditions (a) and (b) imply the conditions of Theorem 6, so they are sufficient for stability.

Necessity: From Theorem 6, (a) is necessary. Furthermore, Theorem 6 implies that there exist integers P_1, P_2, P_1', and P_2' with $P_1P_2' \neq P_1'P_2$ such that $B(\lambda^{P_1}, \lambda^{P_2}) \neq 0$ on $|\lambda| \leqq 1$. Since these are polynomials in positive powers of λ, it follows that they both have no linear phase terms.

Equation (4.25) gives

$$\mathrm{Ind}[B(\lambda^{P_1},\lambda^{P_2})]=P_1\cdot\mathrm{Ind}[B(\lambda,1)]+P_2\cdot\mathrm{Ind}[B(1,\lambda)]=0$$

$$\mathrm{Ind}[B(\lambda^{P'_1},\lambda^{P'_2})]=P'_1\cdot\mathrm{Ind}[B(\lambda,1)]+P'_2\cdot\mathrm{Ind}[B(1,\lambda)]=0\,.$$

Since $P_1P'_2 \neq P'_1P_2$, this implies $\mathrm{Ind}[B(1,\lambda)]=0=\mathrm{Ind}[B(\lambda,1)]$. Of course, the "1" in the theorem statement can be replaced by any complex constant by satisfying $|b|=1$. QED

The conditions of this theorem are similar to those given in Theorem 5.4 which is a first-quadrant stability theorem [4.6, 15]. However, Theorem 7 is much more general because it is true for any type of filter. Hence, no longer is it necessary to check different zero region conditions of $B(w,z)$ for each different type of filter as done in the past [4.3, 8].

Conditions (a) and (b) can be restated, respectively, as (i) the Fourier transform of $b(m,n)$ is never equal to zero, and (ii) there exist no linear phase components in the phase function. These conditions are very similar to restrictions *Dudgeon* [4.23, 12] placed on the Fourier transforms of arrays having rational z transforms to ensure that these sequences have well-defined 2-D complex cepstra. *Dudgeon* also proved that these conditions are equivalent to the phase of the Fourier transform being continuous, odd, and periodic. Here, the phase function [4.1, 23] is defined as the contour integral

$$\phi(u,v)=\mathrm{Im}\,\{\textstyle\int C'_u(z)/C_u(z)dz\}+\phi(u,0)\,, \tag{4.26}$$

where $C_u(z)=\sum\limits_n\Big[\sum\limits_m c(m,n)\exp(-jum)\Big]z^n$, the prime denotes differentiation with respect to z, and the contour integral starts at $z=1$ and proceeds around unit circle to $z=\exp(jv)$. This is called the unwrapped phase of the sequence $c(m,n)$. Other equivalent definitions for the two-dimensional unwrapped phase exist. Theorem 7 along with the above observations suggests the following theorem.

Theorem 8. A recursive filter represented by a recursive filter array is stable iff the unwrapped phase is continuous, odd, and periodic.

Proof. If the recursive filter is stable then by Theorem 7 and a result due to *Dudgeon* [4.1, 23] the unwrapped phase is continuous, odd and periodic. If the phase is defined and continuous, odd, and periodic then $B(w,z)\neq 0$ on T^2 because otherwise the phase would not be defined at the point where $B(w,z)=0$. Moreover, the phase of any projection, i.e., $B(\lambda^{l_1},\lambda^{l_2})$, $l_i\in Z$, must be continuous, odd, and periodic so, in particular $B(\lambda,1)$ and $B(1,\lambda)$ must have continuous, odd, and periodic phases. This, in turn, implies that they have no linear phase component. QED

This theorem is important because it relates the phase of a two-dimensional recursive filter to its stability. This will aid in our development of efficient, practical stability testing algorithms. In the next section, several new stability testing algorithms will be introduced and compared with existing methods.

4.7 Introduction to Stability Tests

The previous sections reviewed and extended many stability theorems. However, these theorems have little practical value if they cannot be translated into numerical methods for testing stability. In the following sections we will develop several new methods for testing the stability of two-dimensional recursive filters based on the previously studied stability theorems. It will be argued that several of these procedures can be implemented more efficiently than existing schemes. Furthermore, several other methods will be reviewed and extended. A unified approach for interpreting many stability algorithms will be introduced. Virtually all the stability tests will be evaluated in terms of algorithmic and computational complexity and in terms of accuracy and numerical effects.

Initially, one may decide to forget about the theorems which relate stability to the zero distribution of $B(w,z)$ and work directly with $h(m,n)$, the impulse response. For instance, the most basic stability theorem for LSI systems is that $\sum_{(m,n)} |h(m,n)| < \infty$. However, checking the absolute summability condition directly is very difficult if not impossible for nontrivial filters. In general, the impulse response of the filter represented by the recursive filter array $b(m,n)$ has no closed form representation. This is contrary to the case in one dimension. In one-dimensional recursive filtering, the system function can be rewritten in the form of a partial fraction expansion from which each factor contributes an exponential term to the impulse response. The Fundamental Theorem of Algebra allows this procedure to be carried out on all one-dimensional systems. However, in two dimensions no Fundamental Theorem of Algebra exists, [4.24] so in most cases $B(w,z)$ cannot be factored. Therefore, this easy method for calculating a closed form impulse response is not applicable.

Another method which may sometimes yield a closed form solution for $h(m,n)$ is to utilize (4.4) the inverse 2-D z transform. However, only the very simplest of systems will yield a closed form representation (see Example 3). On the other hand, even if $h(m,n)$ can be expressed as a function m and n, there is no guarantee that absolute summability could be determined. Again, this is in contrast to one-dimensional filters. In one dimension, if $h(n) \to 0$ as $n \to \infty$, the stability of the recursive filter is assured. However, $h(m,n) \to 0$ as $m, n \to \infty$ does not imply that $h \in l^1$, in fact, it doesn't even imply that $h \in l^2$ [4.10, 25]. Therefore, it must be concluded that stability tests which directly determine if $h \in l^1$ are infeasible for all but the simplest of systems.

4.8 Tests from Stability Theorems

A variety of theorems which related the stability of a recursive filter $b(m,n)$ to the zero region of $B(w,z)$ were presented earlier. In general, most of these theorems have two parts. The first step usually consists of checking the zero distribution of a one-dimensional polynomial as in Theorems 5 and 6. However, in some cases, two one-dimensional polynomials with both positive and negative powers will be required to have a special zero distribution as in Theorem 7. The second step consists of determining whether $B(w,z)$ is zero free on some higher dimensional surface. In Theorems 6 and 7, this surface has two Euclidean dimensions, three in Theorem 5.2, and four in 5.1.

In general, it is much easier to check the one-dimensional condition and hence, this should always be done first. In many cases, this just entails determining whether one one-dimensional polynomial is minimum phase, that is, has all its zeros outside the unit circle. This can be tested by a variety of the methods. However, in practice these polynomials will usually have orders less than twenty, so the Jury table test which will be considered later should be the most efficient procedure. Later, it will be shown that checking the zero distributions of very high order one-dimensional polynomials will be a method to approximately check the multidimensional stability condition for 2-D recursive filter. In these cases, the Jury table will no longer be the most accurate and efficient procedure.

Because these one-dimensional conditions are so easy to check as opposed to the multidimensional conditions, it may be beneficial to test some extra one-dimensional polynomials. The rationale behind this is that if a filter can occasionally be found to be unstable by checking several well-chosen necessary conditions, then in the long run, many calculations may be saved. For instance, for a first-quadrant filter, it will usually be worthwhile to determine the stability of $B(w,1)$, $B(1,z)$, $B(0,z)$, $B(w,0)$, and $B(z,z)$ before proceeding to the multidimensional condition. Furthermore, for a general recursive filter, it may be beneficial to check the stability of the two one-dimensional filters which form the edges of $b(m,n)$, that is,

$$c_1(p)=b(M_1p,N_1p) \quad \text{and} \quad c_2(p)=b(M_2p,N_2p)\,.$$

These must be stable sequences because they completely determine $h(m,n)$ along the (M_1,N_1) and (M_2,N_2) directions.

The major computational effort for testing stability comes from checking whether $B(w,z)$ has zeros in certain regions of C^2 (two-dimensional complex plane). Depending on the type of filter, the critical multidimensional conditions may be

$$B(w,z)\neq 0 \quad \{(w,z):|w|\leqq 1,|z|\leqq 1\}=\bar{U}^2\,, \tag{A}$$

$$B(w,z)\neq 0 \quad \{(w,z):|w|=1,|z|\leqq 1\}=\Delta_{01}\,, \tag{B}$$

$$B(w,z)\neq 0 \quad \{(w,z):|w|=1,|z|=1\}=T^2\,. \tag{C}$$

Recall from Theorems 6 and 7 that the multidimensional condition for any type of filter can be reduced to (C). Since this is only a two-dimensional region while (B) is three and (A) is four, it appears to be the simpler condition to check. Indeed, [4.3] either (B) or (C) is easier to check than (A). However, it is unclear whether condition (B) or (C) is easier to validate. It appears that algorithms which check the weaker condition (C) are just as complicated as those which check (B). In general, (B) need not be true for a stable general recursive filter while (C) must always be valid. However, by Theorem 4 a general recursive filter can always be mapped into a first-quadrant filter and hence, condition (B) must hold for the mapped filter by Theorem 5.3. Therefore, for the remainder of this chapter, we will consider methods which are to be used to directly check the validity of conditions (B) and (C). Furthermore, unless stated otherwise, it will be assumed that the required one-dimensional conditions have already been validated.

The methods for checking (B) and (C) generally fall into two categories. We shall call these the algebraic methods and the approximate methods. In theory, algebraic methods can exactly determine stability in a finite number of steps while approximate methods need an infinite number of steps to acquire that reliability. However, later, we will argue that for all but the simplest of filters, the approximate methods are superior to the algebraic procedures.

In the next section a number of algebraic methods will be studied. We will modify several of the procedures and present a procedure for checking (C) directly. In the following section a variety of approximate methods which will be called the mapping methods for checking (B) will be studied. Several procedures [4.16] will be introduced and some old procedures will be reinterpreted. Most of these methods follow from breaking the two-dimensional condition into a large number of one-dimensional tests. Furthermore, a computationally efficient implementation of one of these procedures which we believe is one of the most efficient methods of checking (B) will be introduced. After this part, several existing stability testing algorithms will be studied. Here, it will be concluded that these algorithms are inferior to some of the previously considered techniques. Next, several approximate methods for testing (C) will be considered. These same methods will also be utilized to form an efficient minimization procedure. Finally, several results pertaining to the frequency response of stable filters will be introduced.

4.9 Algebraic Methods

In theory, there exist stability tests for two-dimensional recursive filters which can precisely determine the stability of any recursive filter. All these tests rely on certain algebraic properties of one-dimensional complex polynomials and therefore they will be called algebraic methods. In the literature [4.3, 7, 13, 26–28] these tests have been developed exclusively for first-quadrant filters which have the form

$$B(w, z) = \sum_{n=0}^{M} \sum_{m=0}^{M} b(m, n) w^m z^n \tag{4.27}$$

with $b(0,0) \neq 0$. Below, we shall review, discuss and add to these results.

As mentioned earlier, *Huang* [4.3] first realized that condition (B) need only be confirmed plus a simple one-dimensional condition to determine stability of a first-quadrant filter. Furthermore, he developed the first algebraic procedure for testing condition (B). Soon after this, a number of other algebraic procedures were developed by other researchers [4.7, 13, 26, 28]. However, all these schemes have the common attribute that they check condition (B).

Most of the algebraic methods can be reduced to two steps. First, a one-dimensional algebraic polynomial root distribution test is applied to the two-dimensional polynomial which is viewed as a one-dimensional polynomial with polynomial coefficients. Secondly, this usually yields real one-dimensional polynomials which must be checked for sign definiteness over a certain real interval. If this step can be validated, then the filter is stable. In some cases this second step can be reduced to or replaced by checking whether a real polynomial has real roots in a given interval or whether a self-inverse polynomial satisfies a specific zero distribution condition.

4.9.1 Huang–Ansell

The first algebraic test to appear in the literature was due to *Huang* [4.3]. Here, the two-dimensional bilinear transform was utilized to transform the system function so that results from multidimensional continuous theory could be employed [4.29]. In effect, instead of testing $B(w,z)\neq 0$ on $\bar{U}^2=\{|w|\leqq 1,\ |z|\leqq 1\}$ the change of variables $s_1=(1-w)/(1+w)$, $s_2=(1-z)/(1+z)$ is performed on $H(w,z)=A(w,z)/B(w,z)$ to obtain $G(s_1,s_2)=E(s_1,s_2)/F(s_1,s_2)$ and $F(s_1,s_2)$ is tested for zeros in $\Gamma=\{(s_1,s_2)\colon \mathrm{Re}\{s_1\}\geqq 0,\ \mathrm{Re}\{s_2\}\geqq 0\}$ using methods developed in [4.29]. Until recently it was believed that if $A(w,z)$ and $B(w,z)$ were mutually prime polynomials then $B(w,z)$ would be free of zeros on $\bar{U}^2$ if and only if $F(s_1,s_2)$ was free of zeros on Γ. Unfortunately, this has been shown to be incorrect [4.30]. Several extra one-dimensional tests must be performed to retain the if and only if condition. However, once these one-dimensional procedures have been performed the Hermite algebraic procedure employed by *Huang* is valid.

This method reduced the stability problem to determining whether a $M\times M$ symmetric polynomial matrix $D(w)$, whose entries are real polynomials of order $2M$, is positive definite for all values of w. *Huang* suggested that this could be verified by showing that the successive principal minors of $D(w)$ are positive for all real w. He further noted that Sturm's method could be used to test positivity of each minor. However, this procedure can be simplified considerably. A result due to *Siljak* [4.26] can be modified to apply to this problem. By altering *Siljak*'s proof to conform to this problem it can be proved that it is not necessary to test all leading minors for positivity. In fact, the following two conditions need be met instead of M:

(i) $D(0)$ be positive definite,
(ii) $\det(D(w))>0$ for all real w.

Hence, this simplification significantly reduces the number of calculations required for this test.

A major difficulty arises in the determination of the polynomial det $[D(w)]$. On a digital computer it is very difficult to compute the coefficients of the determinant of a polynomial matrix in an efficient, programmable manner. We will consider this problem in a later section. However, the one-dimensional s plane procedures of *Routh* and *Frank* [4.31] can be modified so that they give the same information as the Hermite test in a much more programmable manner. In these procedures a series of polynomials are iteratively computed and tested for positivity. These methods have the added advantage that if any positivity tests fail, the rest of the sequence need not be computed. A disadvantage is that the order of the polynomials to be checked for positivity doubles for each row so the last polynomial has order approximately $2^M M$. Hence, even a tenth order filter will present a substantial computational burden. However, for low order filters, this procedure may be more desirable than Huang's Hermite procedure because of its simplicity.

Even with the above simplifications, these tests are not as efficient as some other algebraic methods especially for lower order filters. This is due mainly to the calculation of the bilinear transformation which when efficiently implemented requires of order $O(M^3)$ calculations [4.32]. However, as will be shown later, for large M this is insignificant when compared to the total number of computations needed for any algebraic test.

Other tests can be developed by incorporating different continuous system root distribution procedures. However, most of these s plane methods have analogous z plane procedures which can determine the distribution of zeros with respect to the unit circle. Because these schemes avoid the bilnear transform and because they seem to be a more straightforward way of attacking the problem, most authors have only considered such methods. There are basically three different z plane approaches due to the following authors: *Maria* and *Fahmy* [4.7]; *Anderson* and *Jury* [4.13]; *Siljak* [4.26]; and *Bose* [4.28]. Below, each method will be reviewed briefly and their computational complexity discussed.

4.9.2 The Maria and Fahmy Test

Maria and *Fahmy* [4.7] tested the condition that $B(w,z)\neq 0$ on $|w|=1$ and $|z|\leqq 1$ $[B(w,z)\neq 0$ on $\Delta_{01}]$ by again considering $B(w,z)=\sum_{n=0}^{M} a_n(w)z^n$ where $a_n(w)=\sum_{i=0}^{M} b(i,n)w^i$ and then employing the *Marden–Jury* table [4.33, 34]. This table (see Appendix 4.A) performs basically the same function for determining a complex polynomial's root distribution with respect to the unit circle as the Routh array does for the left-half plane. In fact, a Mth order complex polynomial has all its zeros outside the unit circle if and only if the first column of the table is positive. In the two-dimensional case, the table entries are polynomials in w and w^{-1}. Moreover, the first column entries that are to be checked for positivity are self-inversive polynomials; that is, w_0 is a root so is w_0^{-1}. Two schemes can be employed for checking positivity of such a polynomial. They are *Sturm*'s [4.31] and *Cohn*'s [4.13] methods and they both require approximately $O(N^2)$ operations where N is the order of the polynomial. In order to carry out *Maria* and *Fahmy*'s test completely, the positivity of M polynomials of order M to $2^{M-1}M$ must be checked. Of course, if and of these tests fail, the filter is definitely unstable and positivity of the higher order entries need not be checked. Therefore, it takes less than $O(M^2 4^M)$ real calculations to guarantee that a given Mth order filter is stable. (This number can be reduced significantly if a phase unwrapping procedure is used to determine the root distribution for the high order polynomial.) This means a 16×16 quarter-plane filter stability test requires over a trillion calculations to guarantee stability! This test requires approximately the same number of calculations as the modified s plane procedure based on the Routh array. However, it is slightly easier to implement because no bilinear transform is needed.

4.9.3 The Anderson, Jury, and Siljak Test

Anderson and *Jury* [4.13] tested condition (B) by again considering $B(w,z)$ as a one-dimensional polynomial with polynomial coefficients $a_n(w)$ and then employing the Schur-Cohn test. This as in Huang's procedure concluded in testing a $M \times M$ Hermitian polynomial matrix $C(w)$ with polynomial entries of order $2M$ for positive definiteness on $|w|=1$. *Siljak* [4.26] later showed that this condition can be significantly simplified to checking that $C(1)$ be positive definite and $\det[C(w)]>0$ on $|w|=1$. As discussed earlier, it is difficult to express the coefficients of $\det[C(w)]$ in terms of the coefficients of the polynomial entries of $C(w)$. However, there do exist several efficient, though complicated, methods for computing the exact determinant of matrices with polynomial coefficients [4.35]. In [4.35] several methods were compared and it was concluded that a congruence method was the most efficient. In fact, if the matrix is $n \times n$ and the order of the polynomials is m the total number of calculations needed is approximately $O[mn(mn^2+n^3)]$ for the congruential method while at the same time requiring minimal storage of $O[(m-1)n]$. These results are remarkably lower than other methods such as the Gaussian or expansion by minors methods which, respectively, require $O(m^2n^5)$ and $O(m^2n^2 2^{n-2})$ operations and $O[m(n+1)^3]$ and $O[mn(n^n/2)]$ storage [4.35].

Now, since the Schur-Cohn matrix has polynomial elements of order $2M$ where M is the order of the filter in both w and z, [$C(w)$ is $M \times M$] it requires approximately $O(6M^5)$ operations to obtain the M^2 order polynomial $\det([C(w)]$. Moreover, the positivity can be determined in approximately $O(M^4)$ operations. Therefore, it seems that the Schur-Cohn test when appropriately implemented is more efficient than the previous table test. However, for lower order filters, the table test may be desirable because of its ease of implementation. In addition, Anderson, Jury, and Siljak's procedure cannot terminate early as Maria and Fahmy's test and hence it must always be carried out all the way to determine whether a filter is stable or not. If the congruence method is used in the implementation of Huang's Hermite procedure approximately the same number of calculations are required.

4.9.4 Bose's Test

Schussler [4.36] presented a new stability theorem for one-dimensional polynomials with real coefficients. Later, *Bose* [4.28] extended this technique to implement a new algebraic stability test for two-dimensional filters. The most complicated step of this test calls for determining the resultant $R(x)$ of two $2M$ order polynomials $D_1(x, z)$ and $D_2(x, z)$ derived from $B(w, z)$ and examining it for real roots in the interval $-1 \leqq x \leqq 1$. Unfortunately, it is very difficult to compute $R(x)$. If Sylvester's matrix is employed, $R(x)$ can be found by taking the determinant of $4M$ by $4M$ polynomial matrix [4.37]. There do exist methods where the resultant can be computed using lower order matrices [4.38]. However, the computational burden is never less than the Schur-Cohn algebraic test.

4.9.5 Summary of First-Quadrant Methods

At present, it appears that Anderson and Jury's stability test with Siljak's simplification and with congruential determination of the determinant of the polynomial matrix is the most efficient algebraic stability test for first-quadrant two-dimensional recursive filters. However, it must be stressed that this and any other algebraic methods are infeasible for all but low-order filters. A tenth-order filter requires approximately 10^7 multiply operations with the above method. Moreover, the algorithms that are needed for this test are complicated and difficult to program. Therefore, efficient practical tests need to be developed. Unfortunately, no nonalgebraic technique can guarantee the stability of a given filter. However, most of these methods can be modified to give any desired accuracy.

Although algebraic methods can, in theory, determine the stability of any recursive filter, there exist a number of practical difficulties in their implementation. Since even a low-order filter requires a large number of calculations, a digital computer must be utilized to implement the algorithm. Therefore, the computer's finite word length restriction means that both roundoff error and coefficient quantization effects must be considered. This implies that, in practice, the algebraic methods are no longer exact. Hence, these methods could have difficulties in determining the stability of "critically" stable or unstable filters of which, in theory, can be distinguished. Therefore, only for low-order filters will algebraic methods offer any advantages over some nonalgebraic algorithms.

4.9.6 Examples

In this section closed form stability tests for some simple filters will be catalogued. Some of these results for first-quadrant filters can be found in [4.3, 4] and they are derived directly from these algebraic methods. Furthermore, these results will be generalized by Theorem 4 to apply to other types of filters.

Case 1: $B(w,z)=1+aw$ or $B(w,z)=1+az$.
Then this filter is stable iff $|a|<1$ [4.3]. More generally, by applying Theorem 6.4 one concludes that $B(w,z)=1+aw^nz^m$ is stable iff $|a|<1$ for all integer m and n with $(m,n)\neq(0,0)$.

Case 2: $B(w,z)=1+aw+bz$.
This filter is stable iff $|a|+|b|<1$ [4.3]. With the help of Theorem 6.4 this can be generalized as $B(w,z)=1+aw^{m_1}z^{n_1}+bw^{m_2}z^{n_2}$ is stable iff $|a|+|b|<1$ provided $m_1/n_1\neq m_2/n_2$.

Case 3: $B(w,z)=1+aw+bz+cwz$.
Huang [4.3] determined that $|a+b|-1<c<1-|a-b|$ is necessary and sufficient for stability. Therefore, $B(w,z)=1+aw^{m_1}z^{n_1}+bw^{m_2}z^{n_2}+cw^{m_1+m_2}z^{n_1+n_2}$ is stable iff the same inequalities hold provided $m_1/n_1\neq m_2/n_2$.

Case 4: $B(w,z)=\sum_{m=0}^{2}\sum_{n=0}^{2}b(m,n)w^mz^n$.
It is no longer possible to obtain simple necessary and sufficient conditions directly in terms of the coefficients of $B(w,z)$. Below, we used the Jury table algebraic procedure to obtain several polynomials which must be checked for positivity. These positivity tests can be expressed as a function of the polynomial coefficients. However, since there are so many positivity tests and since they must be modified to handle various special cases we will omit this part and leave it up to the discretion of the user. Now let,

$$
\begin{aligned}
s_2 &= 4(b(0,2)b(2,2)-b(2,0)b(0,0))\\
s_1 &= 2(b(0,2)b(1,2)+b(1,2)b(2,2)-b(0,0)b(1,0)-b(2,0)b(1,0)\\
s_0 &= b^2(0,2)+b^2(1,2)+b^2(2,2)-b^2(0,0)-b^2(1,0)-b^2(2,0)\\
&\quad -2b(0,2)b(2,2)+2b(2,0)b(0,0)\\
t_4 &= b(0,2)b(2,1)-b(2,0)b(0,1)\\
t_3 &= b(0,1)b(2,2)-b(2,1)b(0,0)\\
t_2 &= b(0,2)b(1,1)+b(1,2)b(2,1)-b(1,0)b(0,1)-b(2,0)b(1,1)\\
t_1 &= b(1,2)b(0,1)+b(2,2)b(1,1)-b(0,0)b(1,1)-b(2,1)b(1,0)\\
t_0 &= b(0,2)b(0,1)+b(1,2)b(1,1)+b(2,2)b(2,1)-b(0,1)b(0,0)\\
&\quad -b(1,1)b(1,0)-b(2,1)b(2,0)
\end{aligned}
$$

$$q_4 = t_4 t_3$$
$$q_3 = t_1 t_4 + t_2 t_3$$
$$q_2 = t_0 t_3 + t_1 t_2 + t_4 t_0$$
$$q_1 = t_0 t_1 + t_2 t_0 + t_1 t_3 + t_2 t_4$$
$$q_0 = t_0^2 + t_1^2 + t_2^2 + t_3^2 + t_4^2$$
$$r_0 = s_0^2 - q_0 + 2q_2 - 2q_4$$
$$r_1 = 2s_0 s_1 - 2q_1 + 6q_3$$
$$r_2 = 2s_0 s_2 + s_1^2 + 16q_4 - 4q_2$$
$$r_3 = 2s_2 s_1 - 8q_3$$
$$r_4 = s_2^2 - 16q_4 .$$

Let

$$a_0(z) = b(2,0)z^2 + b(1,0)z + b(0,0)$$
$$b_0(x) = s_2 x^2 + s_1 x + s_0$$
$$c_0(x) = r_4 x^4 + r_3 x^3 + r_2 x^2 + r_1 x + r_0 .$$

Test for stability:

1) $a_0(z)$ must have no root of magnitude less than or equal to one.
2) $b_0(x) < 0$ $|x| \leqq 1$ which reduces to
   i) $s_0 < 0$,
   ii) $b_0(x)$ has no real roots within $|x| \leqq 1$.
3) $c_0(x) > 0$ $|x| \leqq 1$ which reduces to
   i) $r_0 > 0$,
   ii) $c_0(x)$ has no real roots in interval $|x| \leqq 1$.

Either Sturm's test [4.39] or a root finding algorithm can be used to check conditions 2ii) and 3ii). (1), (2), and (3) constitute a set of necessary and sufficient conditions for stability.

4.10 Mapping Tests for First-Quadrant Filters

In this section several stability tests will be developed. Efficient algorithms for implementing those tests will be derived. Moreover, computationally attractive algorithms for existing tests will be developed. This section should unify a large class of efficient stability testing algorithms.

Another approach for checking condition (B), that is, $B(w,z) \neq 0$ on $|w| = 1$, $|z| \leqq 1$, is to effectively map the circle $|w| = 1$ into the z plane according to $B(w,z) = 0$ and determine if the image lies outside the disk $|z| \leqq 1$. As a consequence of algebraic function theory [4.40], these zero sets form continuous curves in the z plane. Therefore, condition (B) is satisfied iff none of those curves intersect the unit disk. If any of these curves do intersect the unit disk, this implies that there exists a (w_1, z_1) such that $|w_1|=1$, $|z_1|\leqq 1$ with $B(w_1, z_1)=0$. Therefore, in theory, the stability of a filter can be determined exactly by checking the zero distribution of the one-dimensional polynomial in z, $B(w_0, z)$ for each fixed w_0 with $|w_0|=1$. Since w has unity magnitude, we will alternately specify these one-dimensional polynomials with a parameter u where $w=\mathrm{e}^{-\mathrm{j}u}$ for $0\leqq u<2\pi$ as

$$B(\mathrm{e}^{-\mathrm{j}u}, z)=\sum_{n=0}^{M}\left[\sum_{m=0}^{M} b(m,n)\mathrm{e}^{-\mathrm{j}um}\right]z^n \tag{4.28}$$

and occasionally we will write $B(\mathrm{e}^{-\mathrm{j}u}, z)$ as $B(u, z)$ and $B_u(z)$ to emphasize that (4.28) should be interpreted as a polynomial in z with coefficients which are functions of u. So,

$$B(u, z)=B_u(z)=B(\mathrm{e}^{-\mathrm{j}u}, z)\,. \tag{4.29}$$

Methods which attempt to obtain stability information about two-dimensional recursive filters by examining $B(u, z)$ for $0\leqq u<2\pi$ will be called mapping methods. We use the name "mapping methods" because these techniques acquire information about the zero maps defined implicitly by the mapping $B(w, z)=0$ for $|w|=1$.

For each fixed u_1, $0\leqq u_1<2\pi$, $B(u_1, z)$ is a one-dimensional polynomial with complex coefficients. As stated above, $B(w, z)\neq 0$ on $|w|=1$, $|z|\leqq 1$ iff $B(u_1, z)$ has no zeros inside or on the unit circle for all u_1 $(0\leqq u_1<2\pi)$, that is, $B(u_1, z)$ must be stable one-dimensional polynomials. Therefore, for each u any root distribution method which can handle polynomials with complex coefficients may be used to check the stability (one-dimensional) of $B(u, z)$. It is important to observe that the roots of $B(u, z)$, and hence the zero maps of $B(w, z)$, do not have to be calculated. In the sections that follow five methods will be studied. They are the root mapping, Nyquist, integrated logarithmic derivative, Jury, and Jensen's methods. It will be discovered that several of these methods give new insights into the stability problem and lend themselves to efficient implementations. In fact, it will be argued that several of these tests when implemented in a certain manner are very efficient methods for testing the stability of most two-dimensional recursive filters.

In practice, the one-dimensional stability of $B(u, z)$ cannot be checked for all u $(0\leqq u<2\pi)$. Only for a finite number of u_j can $B(u_j, z)$ be examined for stability. Unfortunately, once the interval $[0, 2\pi]$ is sampled on a countable grid the accuracy of the above scheme suffers. Indeed, if for a particular u_j one finds, that $B(u_j, z)$ has a zero inside or on the unit circle, then the filter is definitely unstable. However, if $B(u_j, z)$, $j=0,\ldots,L-1$ are all stable, this does not imply that $B(w, z)\neq 0$ on $|w|=1$, $|z|\leqq 1$. Even if we check $B(u_j, z)$ on a countably infinite set stability cannot be guaranteed. For instance, an example can be constructed where the only point in the zero set of $B(w,z)$ that intersects the region $|w|=1$, $|z|\leqq 1$ is (e^{ju_0}, e^{jv_0}) where u_0 is an irrational number. In this case, the above scheme will virtually always fail, unless, by chance, a u_j happens to fall on u_0. Apparently, for this reason, researchers decided to develop algebraic tests which do not have this limitation. However, we plan to argue throughout this section that the above mapping approach is accurate enough and it appears to be the only feasible means for testing the stability of higher order filters.

Although there do exist examples where the mapping method will fail, for the vast majority of filters a fine enough sampling of $[0, 2\pi]$ will yield good results. This follows from [4.34, 40] the fact that the root maps of $B(w,z)=0$ form continuous curves and hence if there exists a (u_0, z_0) with $|z_0|<1$ such that $B(u_0, z_0)=0$ then there exists an interval around u_0 say U_0 such that if $u\in U_0$, $B(u,z)$ will have at least one root inside the unit circle. Therefore, if L, the number of samples on $[0, 2\pi]$ (usually equally spaced) is large enough, then a u_j will fall in the interval U_0 and the test will determine that the filter is unstable. Hence, the number of samples, L, determines the accuracy of the test.

Unfortunately, the larger L is chosen, the more polynomials $B(u_j, z)$ $j=0,\ldots,L-1$ there are that must be checked for stability and hence the more time consuming the test is. However, an efficient means of computing the coefficients of the polynomials $B(u_j, z)$ will be presented. This algorithm is based on the fact that (4.28) can be rewritten as

$$B(e^{-ju}, z)=\sum_{n=0}^{M} C_n(u)z^n\,, \tag{4.30}$$

where

$$C_n(u)=\sum_{m=0}^{M} b(m,n)e^{-jum}\,. \tag{4.31}$$

Equation (4.31) is just the Fourier transform of the rows of the recursive filter array, $b(m,n)$. This suggests that the discrete Fourier transform (DFT) can be employed to compute the coefficients of $B(u_k, z)$ for $u_k=\dfrac{2\pi k}{L}$, $k=0,\ldots,L-1$, so that

$$C_n\left(u=\frac{2\pi k}{L}\right)=\sum_{m=0}^{M} b(m,n)e^{-j2\pi km/L} \tag{4.32}$$

and

$$B(e^{j2\pi k/L}, z)=\sum_{n=0}^{M}\left[\sum_{m=0}^{M} b(m,n)e^{-j2\pi km/L}\right]z^n\,. \tag{4.33}$$

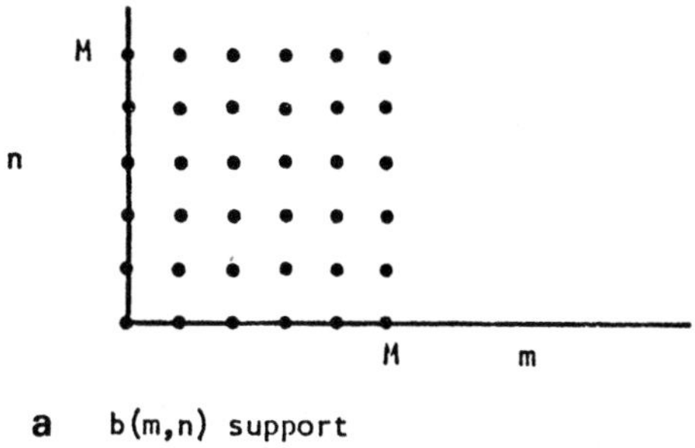

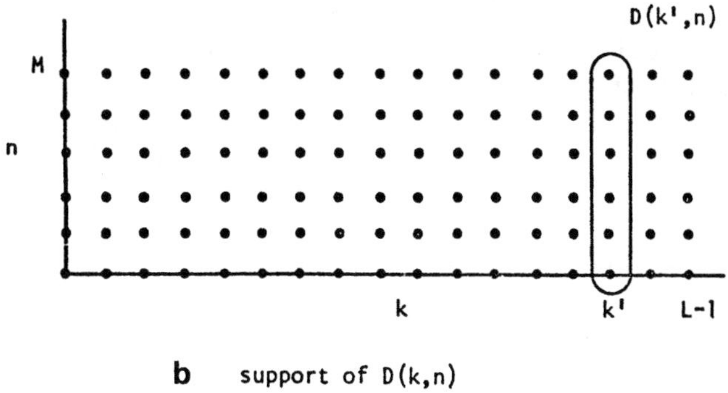

Fig. 4.7a, b. General mapping algorithm for stability. (**a**) $b(m,n)$ support; (**b**) support of $D(k,n)$

For convenience, $C_n\left(\frac{2\pi k}{L}\right)$ and $B(e^{j2\pi k/L}, z)$ will also be written as $C_n(k)$ and $B(k,z)$ or $B_k(z)$, respectively. For appropriate choices of L (usually a power of two), the FFT algorithm can be employed to efficiently compute the DFT. Therefore, an extremely efficient means of obtaining the coefficients of $B(u_k, z)$ for u_k equally spaced is as follows:

1) FFT (length L) each row of $b(m,n)$ to obtain M complex sequences $C_n(k)$ of length L.

2) Form two-dimensional array $D(k,n)$ with $C_n(k)$, $k=0,\ldots,L-1$ as the rows, $D(k,n)$ is an L by $M+1$ complex array.

Next, any one of a number of one-dimensional polynomial root distribution tests can be employed to determine if the one-dimensional sequences forming the columns of $D(k,n)$ are stable. Note that for a fixed k, $D(k,n)$ for $n=0,\ldots,M$ are the coefficients of the polynomial $B\left(u=\frac{2\pi k}{L}, z\right)$. [$D(k,n)$ will also be written as $D_k(n)$.] Therefore, our approach is to first calculate all the needed coefficients of all the one-dimensional polynomials and then check each for stability. This procedure allows the employment of the FFT algorithm and hence is very efficient. Indeed, it is many times more efficient than computing (4.33) for each value of k. Figure 4.7 summarizes this general algorithm. A necessary condition for stability is that for each k', the sequence $D(k',n)$ must be stable.

In the following subsections, a number of the methods which employ the above approach will be evaluated. Furthermore, the efficiency and accuracy of these one-dimensional stability tests when coupled with the above algorithm for computing $D(k,n)$ will be studied. A number of examples and side issues will also be considered.

4.10.1 Root Mapping

The root mapping method is based on the premise that $B(w,z) \neq 0$ on $|w|=1$, $|z| \leqq 1$ iff the root maps of $B(u,z)$ as a function of u do not intersect the unit disk $|z| \leqq 1$. *Huang* [4.3] and *Shanks* [4.4] implemented this test by calculating the minimum magnitude root of $B(u_j, z)$ for a finite number of u_j. These roots are then plotted in the z plane to form a curve. The filter is unstable of this curve intersects the unit disk. This algorithm can be improved immensely by employing our algorithm for obtaining $D(k,n)$. In fact, since the FFT algorithm requires only approximately $L \cdot \log_2 L$ operations this implementation becomes increasingly more efficient as L increases. However, even with this improvement, this approach is rather inefficient because the roots of L polynomials must be computed. [If $b(m,n)$ are real, then by symmetry only $L/2+1$ polynomials need be checked.]

Figures 4.8 and 9 are plots of these minimum norm roots for two examples. In Fig. 4.8,

$$B_1(w,z) = 1 - 1.5w + 0.6w^2 - 1.2z + 1.8wz - 0.72w^2z + 0.5z^2 - 0.75wz^2 + 0.29w^2z^2$$

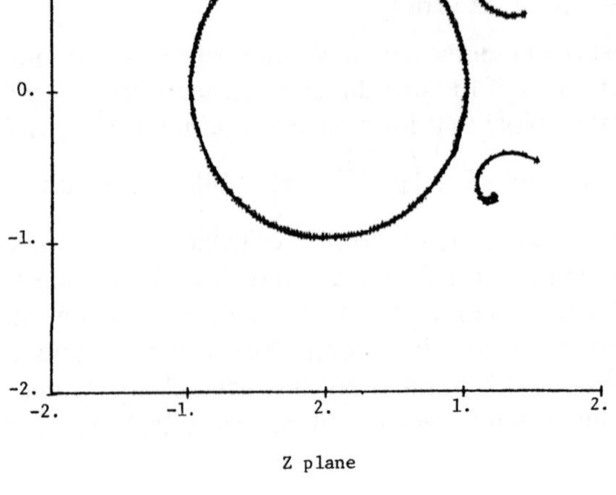

Fig. 4.8. Root map of $B_1(w,z)$

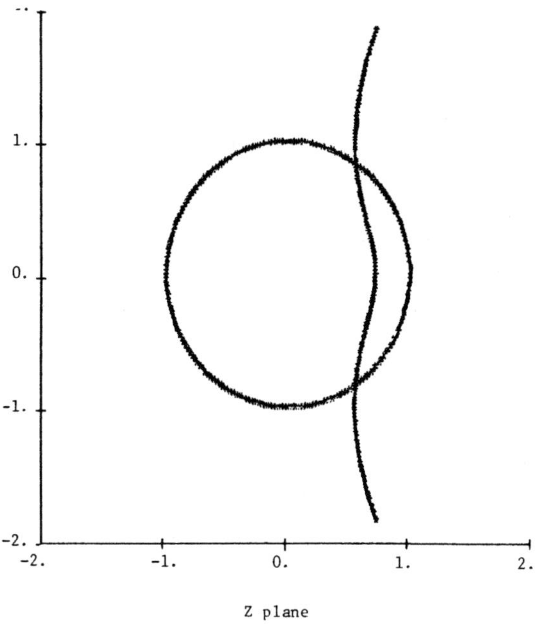

Fig. 4.9. Root map of $B_2(w, z)$

and the plot indicates that $B_1(w, z) \neq 0$ on $|w| = 1$, $|z| \neq 1$. In Fig. 4.9,

$$
\begin{aligned}
B_2(w, z) = {} & 1 - 1.5w + 0.6w^2 - 1.2z + 1.8wz - 0.72w^2z + 0.5z^2 \\
& - 0.75wz^2 + 0.25w^2z^2
\end{aligned}
$$

and the plot indicates that $B(w, z) = 0$ on $|w| = 1$, $|z| \leqq 1$ so it is unstable. These examples also point out how sensitive the root maps can be to the coefficients of $B(w, z)$. B_1 and B_2 only differ by 0.04 in the $b(2, 2)$ term but their root maps differ greatly.

4.10.2 Nyquist Test

Recall from an earlier section that the stability of a one-dimensional polynomial can be determined by means of the Nyquist test. For each u, the curve $B(u, e^{jv})$, $0 \leqq v < 2\pi$ must not encircle nor pass through the origin. This is both a necessary and sufficient condition for $B(w, z)$ to be free of zeros on $|w| = 1$, $|z| \leqq 1$. This same fact was proved recently by *DeCarlo* [4.18, 19] but in a much more difficult way. He relied on covering theory and homotopy theory to obtain this result. In our development, this result follows easily from viewing $B(u, z)$ as a one-dimensional polynomial with parameter u and applying the Argument Principle.

A close study of $B(u, e^{jv})$ will reveal that this is just the two-dimensional Fourier transform of $b(m, n)$. For convenience, it will also be written as $B(u, v)$.

In this notation, the Nyquist test can be interpreted as a procedure which determines if slices of $B(u,v)$ in the v direction are functions which encircle or pass through the origin; that is, $B(w,z)$ is free of zeros in $|w|=1$, $|z|\leqq 1$ iff $\mathrm{Ind}_v[B(u,v)]=0$ for all $0\leqq u<2\pi$.

The above discussion suggests that all of the stability information about $b(m,n)$ is present in its two-dimensional Fourier transform, $B(u,v)$. This should not be surprising because Theorem 6 claims the same. Furthermore, if we recall that the one-dimensional Nyquist criterion can be interpreted using the unwrapped phase function, it follows that this discussion is a demonstration of Theorem 7. [It is assumed here that $B(w,1)$ has no linear phase component along $|w|=1$.] Therefore, $B(w,z)$ is free of zeros on $|w|=1$, $|z|\leqq 1$ iff the unwrapped phase along the v direction can be defined and has no linear phase component for each u.

With the help of this interpretation and [4.16, 41] an efficient algorithm for testing, condition (B) can be developed. First, use our FFT procedure to acquire $D(k,n)$. Next, for each k, unwrap the phase of the sequence $D_k(n)$ using the adaptive phase unwrapping algorithm discussed in [4.16, 41]. [Here, $D_k(n)$ is considered to be a "time" sequence, that is, phase unwrapping gives information about the roots of $\sum_{n=0}^{M} D_k(n)z^n$.] If for any k, the unwrapped phase along the v direction contains a linear phase term, then the filter is definitely unstable. However, if none of the sequences contain a linear phase component, then if L is large "enough" it is "reasonable" to expect that $B(w,z)$ has no zeros on $|w|=1$, $|z|\leqq 1$. As mentioned earlier the accuracy of this test can be enhanced by simple increasing L, the FFT size.

DeCarlo [4.18] implemented his Nyquist test by literally mapping $B(u,e^{jv})$ for each fixed u and looking for encirclements of the origin. However, this method has two shortcomings. First, in light of our FFT method, it is very inefficient to plug u_j into $B(e^{ju},z)$ to obtain the coefficients of z. Secondly, for each u his implementation of the Nyquist test is both inefficient and inaccurate. (It should be stressed that the purpose of his work [4.18] was to develop a Nyquist-like theory for two-dimensional filters and not necessarily derive efficient algorithms for implementing these tests.) The inefficiency problem arises because the FFT algorithm is not used. The accuracy problem arises because the mapping method can lead to ambiguities. This does not occur in the adaptive phase unwrapping algorithm because it has the potential of adapting. It appears that the adaptive phase unwrapping algorithm as described in [4.16, 41] is the most accurate as well as the most efficient for a given accuracy unit circle Nyquist testing procedure. However, since the sequences $D_k(n)$ have length $M+1$ which for most recursive filter applications is less than twenty or so, it follows from that this algorithm is not the most efficient means of extracting zero distribution information. In fact, it has been argued [4.16] that the Jury table is probably one of the most efficient root distribution procedures for such low-order sequences.

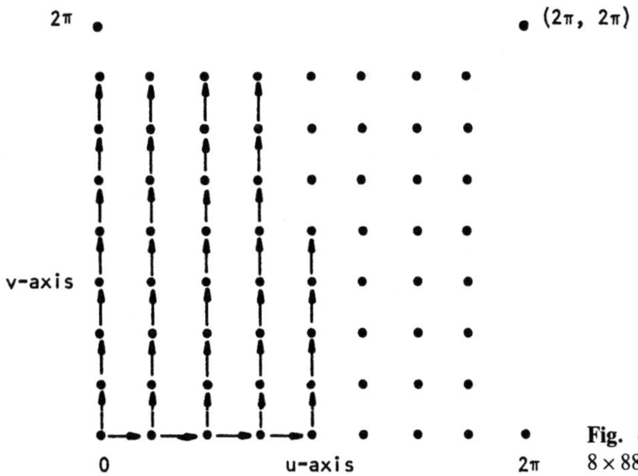

Fig. 4.10. Phase unwrapping on 8×88 grid

One consequence of applying the phase unwrapping procedure to a stable filter is that it almost computes the sampled two-dimensional unwrapped phase. The only reason the above method may not yield the correct sample unwrapped phase is that there may be a phase ambiguity along the u axis. This may allow the phase values along a particular column (v direction, u fixed) to all be off by the same integral multiple of 2π. This ambiguity does not affect the above test because only the presence of a linear phase term need be detected. In order to obtain the correct unwrapped phase, the phase along the u axis must be computed first. This can be easily found by unwrapping the phase of the one-dimensional sequence $c(m) = \sum_{n=0}^{M} b(m,n)$. By unwrapping the phase $c(m)$ we obtain $\theta(k,0)$, the sampled version of the true continuous phase $\theta(u,v)$ along the u axis. While unwrapping the phase of $D_k(n)$ will give $\theta(k,l) + 2\pi a_k$ where a_k is an integer which is a function of k. Combining this information with $\theta(k,0)$ will yield the true sampled unwrapped phase $\theta(k.\,l)$. Figure 4.10 illustrates in which order the phase samples on an 8×8 2D-FFT grid are computed if the above procedure is followed. As shown, only about half the phase samples need be computed because $b(m,n)$ is assumed real. It will be shown later that this two-dimensional phase unwrapping procedure is an important tool for computing 2-D complex cepstra. *Dudgeon* has developed a similar phase unwrapping procedure in order to compute 2-D cepstra.

An interesting consequency of unwrapping the phase of $c(m)$ is that it gives root distribution information about $B(w,1) = \sum_{m=0}^{M} c(m) w^m$. Therefore, by Theorem 5.3 the unwrapped phase contains all the stability information. If it is continuous, odd, and periodic, then no linear phase components exist so the filter is stable and vice versa. Hence, Theorem 8 follows again (at least for first-quadrant filters).

Another interesting result is that if any of the sequences $D_k(n)$ has an unwrapped phase with a linear phase term while $B(w,1)$ and $B(1,z)$ have no such term then the unwrapped phase of $b(m,n)$ is not well defined. This statement is also true if $b(m,n)$ is complex and not a recursive filter array. The phase is ill-defined for two reasons. First, since $B(w,1)$ and $B(1,z)$ have no linear phase term by Theorem 8, the existence of a linear phase term must mean that there exists a (u_0, v_0) $0 \leqq u_0, v_0 < 2\pi$ such that $B(u_0, v_0) = 0$. Therefore, the phase cannot be defined at (u_0, v_0). The phase function cannot be patched up as the one-dimensional case where a zero on the unit circle can be interpreted to mean that the phase jumps by π at that point because it is not known which way the phase jumps. Secondly, if the above algorithm is applied to the columns of $b(m,n)$ so that $D(m,l) = D_l(m)$ phases are unwrapped, we will find that a different unwrapped phase will result. This will be discussed in greater detail in a later section.

4.10.3 Integrated Logarithmic Derivative

Recall that the general approach of this section for checking condition (B) is to determine for each w with $|w| = 1$ if $B(w,z) = 0$ for $|z| \leqq 1$. With the aid of (4.19a), we can interpret this in another way. Equation (4.19a) states that

$$N(w) = \frac{1}{2\pi \mathrm{j}} \int_{|z|=1} \frac{\partial B(w,z)}{\partial z} [B(w,z)]^{-1} dz \,, \tag{4.34}$$

where $N(w)$ for a fixed w is the number of zeros in z of $B(w,z)$ that lie in the region $|z| \leqq 1$. Therefore, $B(w,z) \neq 0$ on $|w| = 1$, $|z| \leqq 1$ iff $N(w)$, considered as a function of $w = \mathrm{e}^{-\mathrm{j}u}$, always equals zero. $N(w)$ can also be interpreted as the winding number of the polynomial $B(w,z)$ for a fixed w. Interpreted in this way, (4.34) implies that no linear phase component of $B_w(z)$ can exist for $B(w,z) \neq 0$ on $|w| = 1$, $|z| \leqq 1$ or that $B_w(z)$ has no zeros inside or on $|z| \leqq 1$ for each w with $|w| = 1$. Therefore, in theory, (4.34) is equivalent to the previously discussed procedures. However, (4.34) can be employed to yield another computational procedure for testing condition (B). Since $w = \mathrm{e}^{-\mathrm{j}u}$ and since the integration is over $z = \mathrm{e}^{-\mathrm{j}v}$ $0 \leqq v < 2\pi$, then

$$N(u) = N(\mathrm{e}^{-\mathrm{j}u}) = \frac{1}{2\pi} \int_{-\pi}^{\pi} \frac{\partial B(u,v)}{\partial v} [B(u,v)]^{-1} dv \,. \tag{4.35}$$

For each fixed u, this method is equivalent to the procedure described in [4.42] for approximately determining the number of zeros outside the unit circle. Therefore, for each sequence $D_k(n)$, $n = 0, \ldots, M$ (4.36) should be evaluated to find the sequence's zero distribution.

$$\tilde{m}_0 = \frac{-1}{jN} \sum_{k=0}^{N-1} \frac{X'(k)}{X(k)} = \frac{1}{N} \frac{\mathrm{FFT}(nX(n))}{\mathrm{FFT}(X(n))} \,. \tag{4.36}$$

However, as discussed in [4.42], there may exist some computational problems with using (4.36) to determine the zero distribution of the polynomial $\sum_{n=0}^{M} D_k(n)z^n$. Therefore, direct evaluation of (4.35) is not recommended. As argued in [4.16], adaptive phase unwrapping is a more efficient and accurate method for acquiring this information.

4.10.4 Jury Test

Instead of employing root calculation, phase unwrapping or (4.35) to determine if $B(u, z)$ has any roots in $|z| \leqq 1$ for $0 \leqq u < 2\pi$, one can use the Jury table (see Appendix 4.A). An efficient implementation of this method is to apply the table procedure to each sequence $D_k(n)$. If any of these sequences are not stable, then the original filter is unstable.

In practice, virtually all two-dimensional recursive filters will have orders less than twenty or so. It appears [4.16] that the above implementation of the Jury test is the most efficient mapping method. Furthermore, it is very easy to program. A more detailed comparison of algorithms will be presented later.

The following examples are tested for stability using three of the previous four algorithms.

Example 8 (E8)

$$b1 = \begin{bmatrix} 1 & 0.5 & 0.25 \\ 0.5 & 0.25 & 0 \\ 0.25 & 0 & 0 \end{bmatrix} \qquad b2 = \begin{bmatrix} 1 & -0.75 & 0.9 \\ 1.5 & -1.2 & 1.3 \\ 1.2 & 0.9 & 0.5 \end{bmatrix}$$

$$b3 = \begin{bmatrix} 1 & -1.2 & 0.5 \\ -1.5 & 1.8 & -0.75 \\ 0.6 & -0.72 & 0.2718 \end{bmatrix} \qquad b4 = \begin{bmatrix} 1.0 & -1.2 & 0.5 \\ -1.5 & 1.8 & -0.75 \\ 0.6 & -0.72 & 0.28 \end{bmatrix}.$$

Filter b1 is very stable, b2 is very unstable, b3 is barely unstable, and b4 is barely stable. Here, degrees of stability is related to how closely the zero curve comes to the region $|w| = 1$, $|z| \leqq 1$ or how much of it is intersected. The root test, phase unwrapping procedure, and Jury table method were applied to each filter. For each example and test the row FFT length L was 1024, so for most practical purposes these tests should be very accurate. The results obtained are listed below.

b1: The sequences $D_k(n)$, $n = 0, 1, 2$, for $k = 0, \ldots, L-1$ are all stable so filter is "almost surey" stable.

b2: $D_0(n)$ is not stable so filter is unstable.

b3: For $k = 49$ to 55 all the methods detect a zero inside $|z| \leqq 1$ for $D_k(n)$ so the filter is unstable.

b4: All $D_k(n)$ sequences $k = 0, \ldots, L-1$ are stable so filter is "almost surely" stable.

It should be stressed that the above algorithms can be applied to many higher order filters with no modification. This is definitely not the case with the algebraic methods.

In the next subsection, another approach for obtaining stability information from $B(u,z)$ is studied. This differs from the last three methods because the winding number (number of roots inside unit circle) is never computed.

4.10.5 Jensen's Formula

If $B(w,z) \neq 0$ on $|w|=1$, $|z| \leq 1$ then for each fixed u

$$I(u)=\frac{1}{2\pi}\int_{\pi}^{\pi}\log|B(u,v)|dv=\log\left|\sum_{m=0}^{M} b(m,0)e^{-jum}\right|=\log|E(u)|\,, \tag{4.37}$$

where $E(u)=\sum_{m=0}^{M} b(m,0)e^{-jum}$. The right-hand side of (4.37) is the Fourier transform of the first row of $b(m,n)$, namely $b(m,0)$, $m=0,\ldots,M$. This follows from Jensen's formula which states that

$$\frac{1}{2\pi}\int_{-\pi}^{\pi}\log|X(e^{ju})|du=\log|X(0)|-\sum_i \log|a_i|+\log|c_i|\,, \tag{4.38}$$

where a_i and c_i are zeros and poles of $X(z)$ inside the unit circle. If (4.37) is not valid for a given u_0, then by (4.38) there exists a (u_0, z_0) with $|z_0|<1$ such that $B(e^{-j\,u_0}, z_0)=0$. Therefore, (4.37) is both necessary and sufficient for $B(w,z) \neq 0$ on $|w|=1$, $|z|<1$. In theory, this is not as general as the previous methods because zeros cannot be located on T^2. However, in practice, since only a finite number of samples can be checked, the other methods essentially have this same limitation.

An efficient algorithm for implementing this method is

1) FFT first row of $b(m,n)$ (length L) to obtain $E(k)=\sum_{m=0}^{M} b(m,0)e^{-j\pi mk/L}$.

2) For each k FFT $D_k(n)$ (length K). This results in a column of 2D-FFT of $b(m,n)$, namely $B(k,l)$, $l=0,\ldots,K-1$.

3) For each k compute $\frac{1}{K}\sum_{l=0}^{K-1}\log|B(k,l)|$ and compare to $\log|E(k)|$.

4) If these quantities differ for any k, then the filter is unstable.

One difficulty with this algorithm is choosing a value for K so that the sum in step 3) approximates the integral in (4.37). This is the same problem that occurs in the integrated logarithmic derivative approach. This problem is enhanced when this algorithm is called to determine if one root of magnitude slightly less than unit is inside $|z| \leq 1$. By (4.38) such a root contributes a very small bias to the integral. Therefore, it is difficult to determine if this small difference is due to this bias or just integration error. However, what makes this method attractive is its numerical simplicity. Since most computers have an FFT algorithm, this method can be programmed in a few minutes.

If $E(u)$ of (4.37) satisfies

$$I_E = \frac{1}{2\pi} \int_{-\pi}^{\pi} \log|E(u)|du = \log|b(0,0)| \tag{4.39}$$

then $B(w,0)$ has no zeros in $|w|<1$. Furthermore, if $B(w,0)$ has no zeros in $|w|<1$, then $I_E = \log|b(0,0)|$. By (4.38) if $B(w,0)$ has zeros in $|w|<1$, then $I_E > \log|b(0,0)|$. If (4.37, 39) hold, then

$$\begin{aligned} \frac{1}{2\pi} \int_{-\pi}^{\pi} I(u)du &= \frac{1}{(2\pi)^2} \int_{-\pi}^{\pi} \int_{-\pi}^{\pi} \log|B(u,v)|dudv \\ &= \frac{1}{2\pi} \int_{-\pi}^{\pi} \log|E(u)|du = \log|b(0,0)| \, . \end{aligned} \tag{4.40}$$

Moreover, if (4.37) does not hold, then

$$\frac{1}{2\pi} \int_{-\pi}^{\pi} I(u)du \neq I_E$$

because there exists an interval U of finite measure such that if $u \in U$, then $I(u) > \log|E(u)|$ while for the other values $I(u) \geqq \log|E(u)|$. Therefore,

$$\frac{1}{(2\pi)^2} \int_{-\pi}^{\pi} \int_{-\pi}^{\pi} \log|B(u,v)|dudv = \log|b(0,0)| \tag{4.41}$$

iff $B(w,z) \neq 0$ on $|w|=1$, $|z|<1$ and $B(w,0) \neq 0$ on $|w|<1$. Equation (4.41) also implies that $B(w,z) \neq 0$ on $|w|<1$, $|z|=1$ and $B(0,z) \neq 0$ on $|z|<1$ because the above development can be repeated for this case. Therefore, (4.41) holds iff $B(w,z) \neq 0$ on $\bar{U}^2 - T^2 = \{(w,z) : |w| \leqq 1, \ |z| \leqq 1\} - \{(w,z) : |w|=1, \ |z|=1\}$. Hence, (4.41) can be used as an approximate stability testing procedure. We use the term "approximate" because this, along with the above algorithm, cannot in general distinguish the case where $B(w,z) \neq 0$ on $\bar{U}^2 - T^2$ while $B(w,z)=0$ on T^2. This is similar to a result given in [4.43].

The two-dimensional FFT can be utilized to approximate (4.41). This is not as efficient nor as accurate as the other algorithm in this subsection. First, the previous algorithm may terminate early thus obviating the entire 2D-FFT calculation which is necessary in (4.41). Secondly, (4.41) may "average out" or obscure instabilities which may have been easily detected using the first method. However, the implementation of (4.41) is slightly simpler because only one comparison is required. Below, this algorithm is applied to filters b1–b4.

Table 4.1. Stability test using integrated logarithmic magnitude

FFT size	b 1	b 2	b 3	b4
8×8	2.9×10^{-3}	0.86	1.2×10^{-2}	1.2×10^{-2}
16×16	4.2×10^{-4}	0.87	1.3×10^{-3}	2.1×10^{-3}
32×32	6.6×10^{-9}	0.87	3.1×10^{-4}	8.6×10^{-5}
64×64	4.1×10^{-7}	0.87	4.8×10^{-5}	3.4×10^{-7}
Stable	Yes	No	?	Yes

Since $b(0,0)=1$ in each example, this implies that if the filters are stable, then (4.41) should equal zero. These results are tabulated in Table 4.1. Example b3 is a troublesome case because its zero sheet barely intersects U^2. Here, the numerical difficulties that were discussed earlier become apparent.

4.10.6 Discussion

Figure 4.11 summarizes the mapping algorithms which are employed to determine the stability of first-quadrant filters. In essence, these procedures attempt to determine if $B(w,z)$ is free of zeros on $\bar{U}^2$. In particular, these algorithms try to show that $B(w,z)\neq 0$ on $|w|=1$, $|z|\leqq 1$ and $B(w,0)\neq 0$ on $|w|\leqq 1$ [or $B(w,z)\neq 0$ on $|w|\leqq 1$, $|z|=1$ and $B(0,z)\neq 0$ on $|z|\leqq 1$] because these are necessary and sufficient conditions for $B(w,z)\neq 0$ on $\bar{U}^2$. All these methods rely on the parameterization $w=e^{-ju}$ and try to check for each u if $B(u,z)=0$ on $|z|\leqq 1$. In theory, most of these are exact tests. However, in practice, two main problems occur:

i) Only a finite number of u can be chosen.

ii) Algorithms which determine if $B(u,z)$ is stable may not be exact.

This second difficulty is quite apparent when Jensen's formula or integrated logarithmic derivative method is employed. This is virtually no problem when root calculation, Jury test, and Nyquist test (implemented by adaptive phase unwrapping) are used. However, even with these methods, errors may occur when implemented on a finite precision computer. If one had an infinite precision machine, then only the Jury test method can guarantee the elimination of the problem in a finite number of steps. Problem i) is the main limitation of the algorithms because not even an infinite precision machine can overcome it. However, the accuracy can be increased to almost any desired level by increasing the sampling rate. Efficient implementations of the above algorithms must employ the FFT algorithm to compute these samples.

Although algebraic methods are not affected by limitation i), numerical errors due to finite precision arithmetic and coefficient quantization may be severe. Furthermore, for filters of size $M>4$, these techniques are extremely difficult to program. These methods also require an excessive amount of calculations for even moderate order filters. On the other hand, the mapping

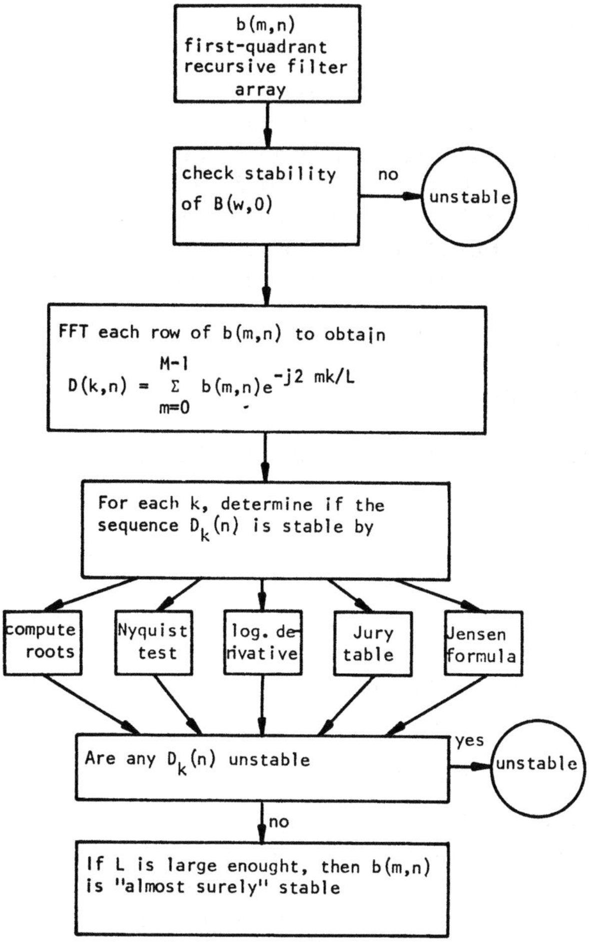

Fig. 4.11. Block diagram summarizing mapping algorithms

algorithms are easy to program and are able to handle virtually any order filter without any modifications. For these reasons we believe that the mapping algorithms are superior to the algebraic techniques. In particular, the Jury table scheme appears to be the most efficient and accurate approach for lower order filters ($M < 20$ or so) while adaptive phase unwrapping should be more accurate and sometimes more efficient for higher order filters.

In the case where it is absolutely necessary to determine if a given filter is stable, it may be argued that approximate methods should not be used. However, a trick can be employed to overcome part of this objection. Let a_1 and a_2 be two real numbers with magnitudes slightly greater than one. If an approximate test with the appropriate accuracy is applied to $B(a_1, w, a_2 z)$ and it is found to be stable, then it is virtually certain that $B(w, z)$ is stable. This is equivalent to checking the stability of the filter $c(m, n) = a_1^m a_2^n b(m, n)$.

4.11 Other First-Quadrant Stability Tests

In this section several other tests will be considered. Several of these have appeared in [4.8, 38]. The main contention of this discussion will be to show that these methods supply stability information in a rather inaccurate and inefficient manner. However, this fact (which we have reached by hindsight) by no means belittles the pioneering works of *Ekstrom* and *Woods* [4.8, 23, 38, 45] in relating stability to cepstra. First, as background, a brief review of the two-dimensional complex cepstrum will be given [4.12]. Then several cepstral [4.8] stability test will be studied.

4.11.1 Two-Dimensional Cepstra

Let $s(m,n)$ be a finite extent signal whose z transform is defined by

$$S(w,z)=\sum_m \sum_n s(m,n)w^m z^n \,. \tag{4.42}$$

The complex cepstrum of $s(m,n)$ is defined as

$$\hat{s}(m,n)=Z^{-1}[\log[Z[s(m,n)]]] \,, \tag{4.43}$$

where $Z[\cdot]$ is the z transform operator and $Z^{-1}[\cdot]$ is its inverse. In order to define the cepstrum properly the function

$$S(w,z)=\log[S(w,z)]=\log[Z[s(m,n)]] \tag{4.44}$$

must be a valid z transform [4.12]. In general, only stable signals with stable cepstra are of interest; that is, both $S(w,z)$ and $S(w,z)$ must have regions of convergence including $T^2=\{(w,z):|w|=1,\ |z|=1\}$. This implies $S(e^{-ju},e^{-jv})\neq 0$ for $-\pi\leqq u,v<\pi$ and that the phase be continuous and periodic [4.23].

There exist two methods for computing complex cepstra. Since T^2 must be in the region of convergence the two-dimensional Fourier transform can replace the two-dimensional (2D) z transform so

$$\hat{s}(m,n)=FT^{-1}[\log[FT(s(m,n))]] \,. \tag{4.45}$$

Another method which can be used for a special class of signals by applying recursive equations can be developed, see [4.12] for further details. During the calculation of $\hat{s}(m,n)$ using (4.45) the following must be computed

$$\hat{S}(u,v)=\log[S(u,v)]=\log|S(u,v)|+j\theta(u,v) \,, \tag{4.46}$$

where $\theta(u,v)$ is the phase of the Fourier transform of the signal. In order for $\hat{S}(w,z)$ to be analytic on T^2, θ must be continuous and periodic. Therefore, an inverse tangent routine cannot be used in its computation because it only supplies the principal value of the phase. A two-dimensional phase unwrapping algorithm must be employed to compute the true continuous phase. Table 4.2

Table 4.2. Complex cepstrum examples

Index (m, n)	$\hat{b}_1(m, n)$ true	$\hat{b}_{1p}(m, n)$ aliased	$\hat{b}_4(m, n)$ true	$\hat{b}_{4p}(m, n)$ aliased
(0, 0)	0	0	0	0.00009
(0, 1)	0.5	0.5	−1.2	−1.20001
(2, 0)	0.125	0.125	−2.2	0.22013
(3, 0)	−0.0833	−0.083333	0.024	0.02388
(4, 0)	0.015625	0.015625	0.0766	0.076515
(1, 1)	0.0	0.0	0.0	−0.000091
(1, 2)	−0.125	−0.125	0.0	−0.000012
(0, 1)	0.5	0.5	−1.5	1.50006
(0, 2)	0.125	0.125	−0.525	−0.52504
(0, 3)	−0.0833	−0.083333	−0.225	−0.22505

FFT size 32 × 32, 6 decimal accuracy

tabulates the results of taking the complex cepstrum of array b1 and b4 using an adaptive phase unwrapping algorithm developed in [4.12, 44]. For these examples the true cepstra $\hat{b}(m,n)$ can be found using an iterative procedure [4.12] so they can be compared to the cepstra obtained by using our algorithm. Since the 2-D DFT is used to calculate the Fourier transform, the phase unwrapping algorithm actually calculates an aliased version of $\hat{s}(m,n)$, [4.12] namely,

$$\hat{s}_p(m,n) = \sum_{i=-\infty}^{\infty} \sum_{j=-\infty}^{\infty} \hat{s}(m+iL, n+jL) \,, \tag{4.47}$$

where $L \times L$ are the dimensions of the 2D DFT. In example b1 for $L=32$, this aliasing is quite small and the correct cepstrum (to at least six decimal places) is computed. However, if the algorithm is used to enumerate the complex cepstrum of b4 with $L=32$, aliasing results. This is the case because $B_4(w,z)$ almost equals zero on T^2 (see Table 4.2).

4.11.2 Cepstral Stability Methods

The basic premise underlying these algorithms is that a general recursive digital filter is stable if and only if its two-dimensional complex cepstrum exists. Furthermore, the complex cepstrum must be supported in the same minimum angle sector β^* (see discussion in Sect. 4.3 and [4.23, 45, 8]). Therefore, one possible algorithm for determining stability is to use the complex cepstral algorithm presented above. If the filter is stable by Theorem 8 it satisfies the conditions for the existence of complex cepstrum. Therefore, if the FFT size is large enough to minimize aliasing, the result will correspond closely to the true cepstrum which has support on β^*. However, if the filter is unstable then the calculated complex cepstrum will not correspond to the true cepstrum and it will usually have support outside of β^*. Therefore, if the complex cepstrum does not have support on β^*, then the filter is definitely unstable. However, because of aliasing it is sometimes difficult to say with certainty whether the computed cepstrum is identically zero outside of β^* [4.12].

For simplicitly, assume $B(w,1)$ and $B(1,z)$ have no linear phase component and that $B(w,z)=0$ in U^2. By an argument similar to that given in Sect. 4.10.2 the two-dimensional phase function cannot be made continuous and periodic. Two different phase unwrapping algorithms will usually give different results (e.g., unwrapping on columns vs rows). Therefore, computing the complex cepstrum in this case is senseless because the two-dimensional phase cannot be unambiguously defined. In fact, all the stability information is present in the two-dimensional unwrapped phase. If it cannot be defined to be continuous and periodic and thus consistent then the filter is definitely unstable. If it can be shown that the phase is continuous and periodic then the filter is stable. But most of the methods presented in Sect. 4.10 attempt to acquire this information for first-quadrant filters in a simple and efficient manner and thus should be employed instead of the complex cepstrum algorithm. Moreover, a general recursive filter can always be mapped into a first-quadrant filter so again the Sect. 4.10 methods can be used to determine if it has a continuous and periodic phase and thus is stable. In a later section a method will be presented which can be utilized to determine if a continuous and periodic phase and hence a stable complex cepstrum exists given any arbitrary finite extent array $s(m,n)$ (that is, it need not be a recursive filter array).

The original cepstrum stability test [4.8] of which the above method is an extension is based on the following premises: If a filter $b(m,n)$ is stable then its complex cepstrum exists and is restricted to β^*; since β^* has angle less than π then $\hat{b}(m,n)$ is completely specified by its even part $\hat{b}_e(m,n)$ $=\frac{1}{2}[\hat{b}(m,n)+\hat{b}(-m,-n)]$ (this can be shown easily by generalizing the one-dimensional case in [Ref. 4.42, Chap. 7] also see [4.38]); since $b(m,n)$ is a real array

$$\begin{aligned}\hat{b}_e(m,n) &= FT^{-1}[\log|B(u,v)|]\\ &= \frac{1}{(2\pi)^2}\int_{-\pi}^{\pi}\int \log|B(u,v)|e^{jum}e^{jvn}du dv\,. \end{aligned}\tag{4.48}$$

The following algorithm is due to *Ekstrom* and *Woods* [4.8] and follows from the above arguments. It will be called the cepstral stability algorithm.

1) Given $b(m,n)$ a recursive filter array
2) Calculate $|B(u,v)|^2$ via DFT
3) Computer $\log|B(u,v)|^2$
4) IDFT $[\log|B(u,v)|^2]$
5) Window β^* call $\hat{b}'$
6) DFT $(\hat{b})$
7) $\mathrm{Exp}[\mathrm{DFT}(\hat{b})]=B'(u,v)$
8) $\mathrm{IDFT}[B'(u,v)]=b'(m,n)$ and compare to $b(m,n)$ if equal then $b(m,n)$ is stable.

Step 4) is approximately an implementation of (4.48) which can be rewritten as

$$\hat{b}_e(m,n) \simeq \frac{1}{L^2} \sum_{l=0}^{L-1} \sum_{k=0}^{L-1} \log|B(k,l)| e^{j2\pi mk/L} e^{j2\pi nl/L} . \tag{4.49}$$

Note that for $(m,n)=(0,0)$ (4.48, 49) become

$$\hat{b}_e(0,0) = \frac{1}{(2\pi)^2} \int_{-\pi}^{\pi} \int_{-\pi}^{\pi} \log|B(u,v)| du dv$$

$$\simeq \frac{1}{L^2} \sum_{l=0}^{L-1} \sum_{k=0}^{L-1} \log|B(k,l)| . \tag{4.50}$$

This is the same integral given in (4.41) and in Sect. 4.10.5 it was shown that if it equaled $\log|b(0,0)|$ then the filter is stable. (Actually, this was shown for a first-quadrant array but it can easily be shown by applying the mapping theorem in Sect. 4.4 that it is also true for any type of recursive filter array.) Therefore, $\hat{b}_e(0,0)$ holds all the stability information and no other information is needed. Hence, the cepstrum stability algorithm obfuscates the true stability carrying component. Recall that it was concluded in Sect. 4.10.5 that (4.50) is not the most efficient means for checking stability but it is probably the easiest algorithm to implement.

Example 9 (E9)

The complex cepstral and cepstral methods will be applied to examples b1 to b4. For the complex cepstral algorithm the maximum non-first-quadrant point will be given. As mentioned earlier this gives an indication whether or not the filter is stable. In fact, the greater its absolute value the greater the probability that the filter is not stable. For the cepstral algorithm the truncated $b'(m,n)$ will be tabulated along with the maximum value of $b'(m,n)$ outside of this truncation window (3×3 for these examples). In all examples the DFT size is 32×32 and results indicated to six decimal places.

b1: (a) CEP(cepstrum) – max-value outside this array is 0.0001

$$b'(m,n) = \begin{bmatrix} 1.0 & 0.5 & 0.25 \\ 0.5 & 0.25 & 0 \\ 0.25 & 0 & 0 \end{bmatrix} .$$

(b) CCEP (complex cepstrum) – maximum non-first-quadrant point $= 0.0001$.

b2: (a) CEP – 0.1011

$$b'(m,n)=\begin{bmatrix} 2.388 & -0.4687 & 0.6937 \\ 1.1257 & -0.3333 & 0.8118 \\ 0.1331 & 0.3977 & 0.2189 \end{bmatrix}.$$

(b) CCEP – global linear phase component in both m and n directions which implies that $B(w,1)$ and $B(1,z)$ have zeros inside unit circle, hence $b(m,n)$ is unstable.

b3: (a) CEP – 0.0035 ($b(2,2)=0.271$)

$$b'(m,n)=\begin{bmatrix} 0.9995 & -1.2003 & 0.4990 \\ -1.5002 & 1.8007 & -0.7501 \\ 0.6001 & -0.7203 & 0.27117 \end{bmatrix}.$$

(b) CCEP = 0.136.

b4: (a) CEP – 0.0026

$$b'(m,n)=\begin{bmatrix} 0.9999 & -1.2000 & 0.50001 \\ -1.50004 & 1.80001 & -0.75005 \\ 0.60004 & -0.72006 & 0.28002 \end{bmatrix}.$$

(b) CCEP – 0.0037.

Example b4 should be compared with b3. These arrays only differ in the (2, 2) entry by 0.009. This alteration causes b3 to become stable. Notice that the output array b3′(m, n) is similar to b3(m, n). In fact, they are so close that it is difficult to determine whether either of them are stable or unstable. On the other hand, the CCEP algorithm gives a non-first-quadrant component as large as 0.13 thus indicating that the filter is unstable. However, this difference appears because the unwrapped phase cannot be defined to be continuous and periodic. Therefore, as mentioned earlier since all the stability information is in the unwrapped phase the complex cepstrum need not be calculated. In conclusion, the complex cepstrum method appears more sensitive to critical unstable filters than the cepstral method. However, they are both inefficient and inaccurate and should not be used to test stability.

In the next section we will present methods which can determine if $B(w,z)=0$ on T^2 for any $b(m,n)$ (not necessarily a recursive filter array). We will find that the root distribution tests (Sect. 4.10) can again be employed to efficiently implement these algorithms.

4.12 Determination of the Existence of Zeros of $B(w, z)$ on T^2

The techniques developed in this section can be used directly to determine whether or not $B(w,z)=0$ on T^2 and hence check if condition (a) of Theorem 7 is satisfied. These methods are based on the fact that in most cases the existence of zero on T^2 can be determined without actually finding this point. This important observation can be demonstrated by studying the zero region of $B(w,z)$, that is, $\{(w,z):B(w,z)=0,\ (w,z)\in C^2\}$. This region consists of a finite number of two-dimensional "sheets" in C^2 [4.22] where C^2 is the two-dimensional complex plane.

In most cases if a zero sheet intersects T^2 at a point (w_0, z_0) it then intersects $D_{01}=\{(w,z):|w|=1,\quad |z|<1\}$ and $D_{0-1}=\{(w,z):|w|=1,\quad |z|>1\}$ (also $D_{10}=\{(w,z):|w|<1,\ |z|=1\}$ and $D_{-10}=\{(w,z):|w|>1,\ |z|=1\}$) in some neighborhood of (w_0, z_0). Roughly speaking, this is a consequence of viewing the solutions of $B(w,z)=0$ as algebraic functions and applying the Implicit Function Theorem [4.46]. For instance, Fig. 4.12 depicts the solution of this equation in the neighborhood of (w_0, z_0). Case A represents the above observation while case B describes the rarely occurring event where the zero sheet does not intersect both D_{0-1} and D_{01} in the neighborhood of (w_0, z_0).

If case A occurs, the solution of $B(w,z)=0$ in the neighborhood of (w_0, z_0) and in regions D_{01} and D_{0-1} is a one-dimensional curve. If $|w|=1$ is

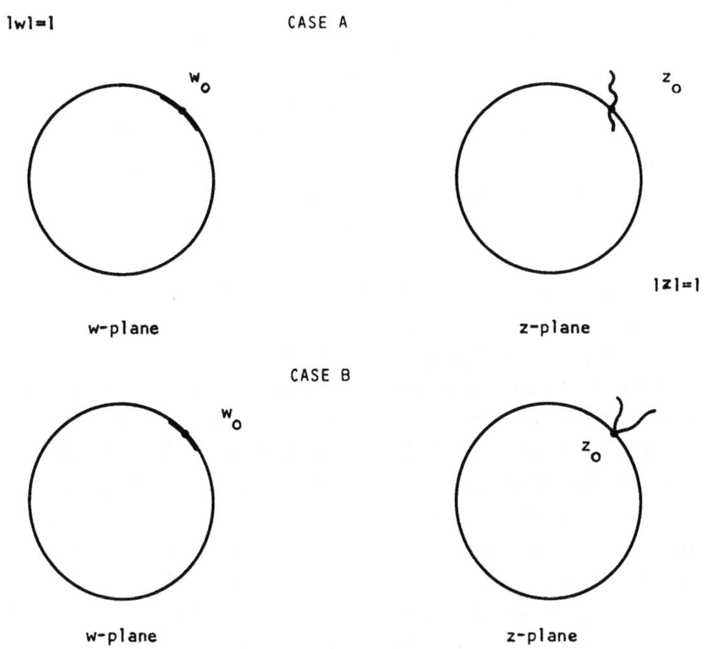

Fig. 4.12a, b. Several possible solutions of $B(w,z)=0$ in the neighborhood of a point $(w_0, z)\in T^2$. **(a)** Case A; **(b)** case B

parameterized as $w=e^{-ju}$ $(0 \leqq u < 2\pi)$ with $w_0 = e^{-ju_0}$ then this curve is one of the root maps of

$$B_u(z) = \sum_n \left(\sum_m b(m,n) e^{-jum} \right) z^n \tag{4.51}$$

which passes through $(w_0, z_0) = (e^{-ju_0}, e^{-jv_0})$. Therefore, there exist two open intervals $U_{10} = (u_1, u_0)$ and $U_{02} = (u_0, u_2)$ such that if $u_a \in U_{10}$ and $u_b \in U_{02}$ then $B_{u_a}(z)$ and $B_{u_b}(z)$ have a different number of zeros inside (outside) the unit circle. This same argument can be applied to $B_v(w)$ where

$$B_v(w) = \sum_m \left(\sum_n b(m,n) e^{-jvn} \right) w^m . \tag{4.52}$$

Conversely, a continuity argument can be employed to show that if the number of roots of $B_u(z)$ or $B_v(w)$ inside the unit circle vary as a function of u or v, respectively, then this implies there exists a $(w_0, z_0) \in T^2$ such that $B(w_0, z_0) = 0$. Thus if $0 \leqq u \leqq \pi$ (or $0 \leqq v \leqq \pi$) is sampled fine enough so there exists a sample in U_{10} and in U_{02} (or V_{10} and V_{02}) then the existence of the point $(w_0, z_0) = (e^{-ju_0}, e^{-jv_0})$ is substantiated although the values for u_0 and v_0 are not explicitly known.

The above discussion suggests the following practical tests for condition (a) of Theorem 7. First, sample either $0 \leqq u \leqq \pi$ or $0 \leqq v \leqq \pi$ or both on a fine grid. The intersample distance will determine the accuracy of the test. Next, evaluate the coefficients of $B_u(z)$ or $B_v(w)$ for all grid samples. This can be done quite efficiently by employing the FFT algorithm on the rows of $b(m,n)$ to obtain the coefficients of $B_u(z)$ for $u_i = 2\pi i/L$ and on the columns of $b(m,n)$ to obtain the coefficients of $B_v(w)$ for $v_i = 2\pi i/L$ (see Sect. 4.10). Next, some method is used to determine the root distribution of each $B_{u_i}(z)$ or $B_{v_i}(w)$ with respect to the unit circle. These polynomials are different than those in Sect. 4.9 because here they can have both positive and negative exponents. For virtually all digital filter applications these polynomials will have orders less than one-hundred and in most cases less than ten. Thus, this implies that the Marden–Jury table method should be one of the most efficient methods for determining this.

However, either phase unwrapping or the integrated logarithmic derivative can also be used here. Note, that Jensen's formula cannot be employed because it does not give root distribution information. [However, the extension of Jensen's formula will be valid, i.e., the double integral of the log magnitude of the Fourier transform of $b(m,n)$ should still be zero if the filter is stable.] At any rate, the table method should be the most efficient. The last step of this procedure is to check whether this zero distribution ever changes as a function of u_i or v_i. If it does change then the filter is definitely unstable. On the other hand, if it doesn't change then the filter may or may not be stable. However, if a very dense sampling grid is used then we can be "almost sure" that the filter is stable.

A close study of Sect. 4.10.2 and (4.51, 52) will reveal that they can be used to define the two-dimensional unwrapped phase when substituted into (4.26).

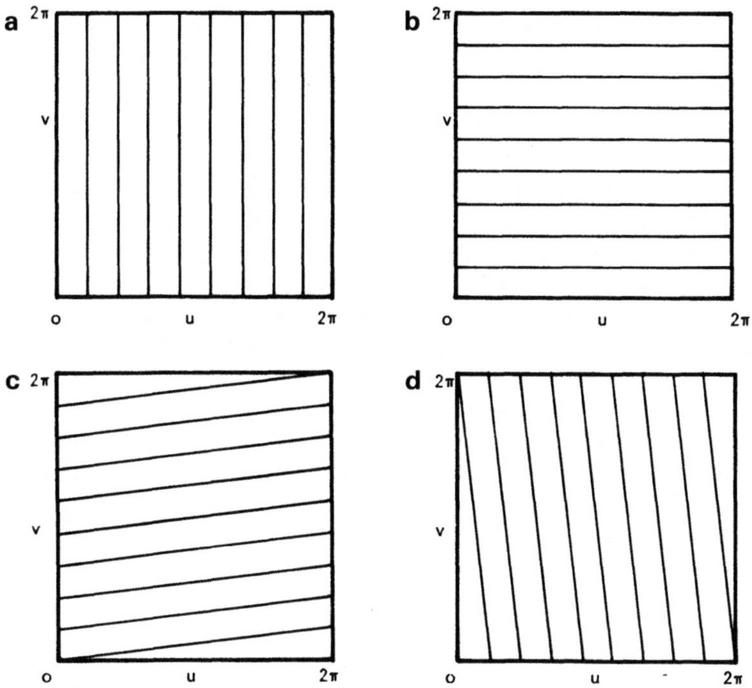

Fig. 4.13a–d. Parameterization of torus T^2 using different projections. (**a**) Row; (**b**) column; (**c**) row concatenation; (**d**) column concatentation

Equation (4.51) defines the phase on a series of vertical lines in the $u-v$ plane while (4.52) allows the phase to be computed on a set of horizontal lines (see Fig. 4.13a, b). If along any of these lines the phase contains a linear component then by the above discussion and Theorem 8 the filter is unstable (and, in fact, the two-dimensional phase is not well defined). As mentioned earlier, it is more efficient to obtain this information via an analytic method such as Marden–Jury table than actually unwrapping the phase along each line.

The above tests require that a number of one-dimensional FFTs be taken; more precisely, one for each column or row. Furthermore, a series of root distribution tests must then be executed. However, we can derive an interesting procedure which requires only one root distribution test with at most one FFT computation. This method was motivated by *Mersereau*'s One-Projection Theorem and the Projection-Slice Theorem [4.47, 48]. In essence, it relies on the fact that if a two-dimensional sequence has a phase which is continuous, odd, and periodic, then any projection of this array has a phase which is also continuous, odd, and periodic; and it is equal to some slice of the two-dimensional phase (see also Theorem 8). By decreasing the angle of the projection, the resulting phase can be calculated on as fine a grid as desired. Figure 4.13c, d show over which values of the $u-v$ plane the phase is calculated for two different projections. In this case, the toroid T^2 is approximately parameterized as a tightly wound helix as opposed to circles which result if (4.51) or (4.52) are used. There exist a number of different projections which will give a desired accuracy. Here, for simplicity, we only consider two. First, the phase of the sequence formed by concatenating the zero padded columns will equal the two-dimensional phase calculated on a slice of angle $\tan^{-1}(1/N)$ depicted in Fig. 4.13c where N is the length of the zero padded columns [4.47, 48]. Moreover, by concatenating the zero padded rows of length M, the above procedure will yield a slice of angle $\tan^{-1}(1/M)$ from the vertical as shown in Fig. 4.13d. The grid on which the slice is computed can be made finer and finer, thus increasing the accuracy of the stability test by adding more zeros to the original array's columns or rows.

Since only the existence of a linear phase term or equivalently, the appropriate zero distribution with respect to the unit circle of this one-dimensional sequence need be determined when either an analytic method or a phase unwrapping algorithm can be employed to extract this information. Generally, this one-dimensional sequence will have several hundred or more elements. It has been our experience that for such large sequences, a modified version of Tribolet's phase unwrapping algorithm [4.41, 16] works faster and more accurately than any other procedure.

So far no mention has been made about the relation of the zero distribution with respect to the unit circle of the projection to that required by continuity of the two-dimensional phase. The one-dimensional sequence associated with column concatenation has a representation of $B(\lambda^N, \lambda)$ while row concatenation has one of $B(\lambda, \lambda^M)$. Note, that these sequences can have both positive and negative indices. By (4.25) and Theorem 6 if $b(m,n)$ is stable then for any integers M and N

$$\mathrm{Ind}\,[B(\lambda^N, \lambda)] = \mathrm{Ind}\,[B(\lambda, \lambda^M)] = 0\,. \tag{4.53}$$

However if $b(m,n)$ is not stable and if

$$\mathrm{Ind}\,[B(\lambda, 1)] = \mathrm{Ind}\,[B(1, \lambda)] = 0$$

then in most cases if N or M is large enough in magnitude we will have

$$\mathrm{Ind}\,[B(\lambda^N, \lambda)] \neq 0 \tag{4.54}$$

and/or

$$\mathrm{Ind}\,[B(\lambda, \lambda^M)] \neq 0\,.$$

Therefore, if for any integers M or N (4.53) does not hold then $b(m,n)$ is unstable. Moreover, if (4.54) holds for very large valves of M or N then we can be "almost sure" that $b(m,n)$ is stable.

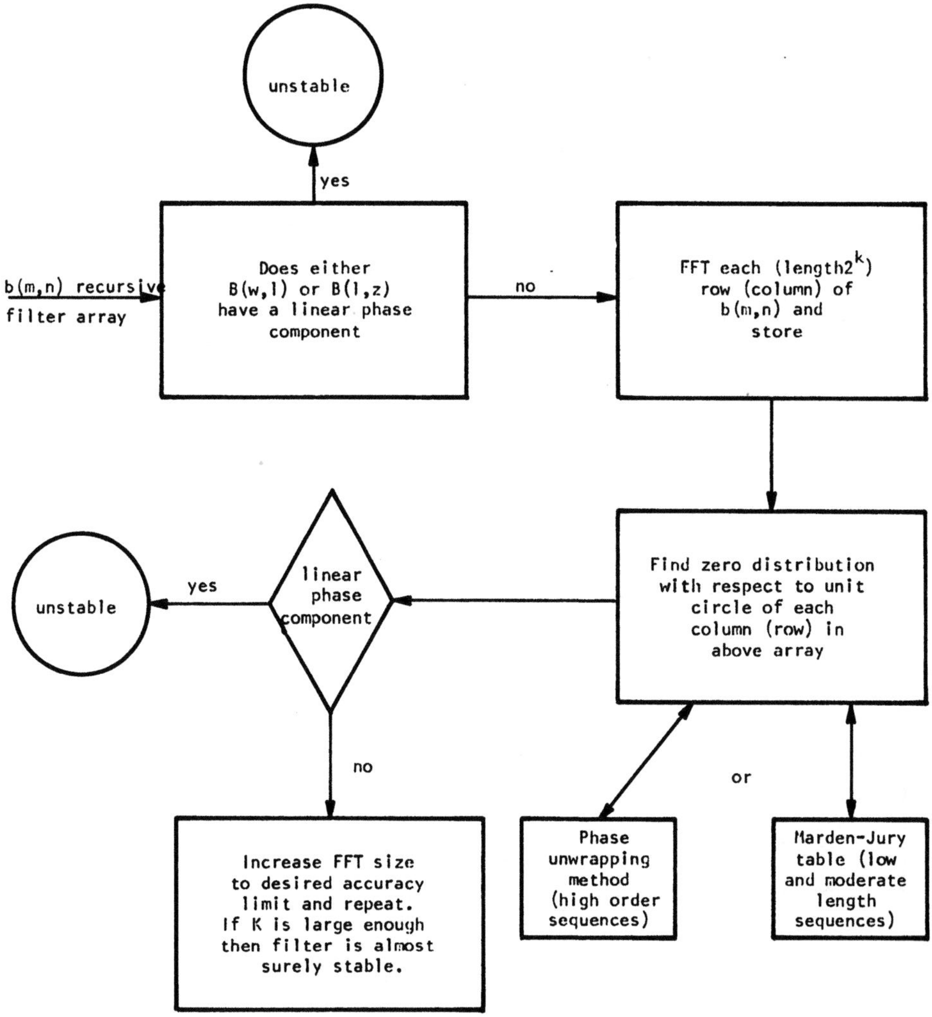

Fig. 4.14. Efficient stability test algorithms for two-dimensional recursive filters (row and column algorithms)

Notice that these row and column concatenation stability tests like the row and column algorithms presented earlier cannot guarantee the stability of a filter. However, the accuracy of these tests can be increased to almost any desired level. On the other hand, there still will exist examples which are unstable but will be indicated as stable by these tests. For example, a zero sheet may just touch T^2 at one point, say $[\exp(-ju_0), \exp(-jv_0)]$ where u_0 and v_0 are irrational numbers, and not pass between D_{01} and D_{0-1}. However, in critical cases as this, the more accurate (at least theoretically) algebraic test when implemented on a computer with its finite precision arithmetic may also

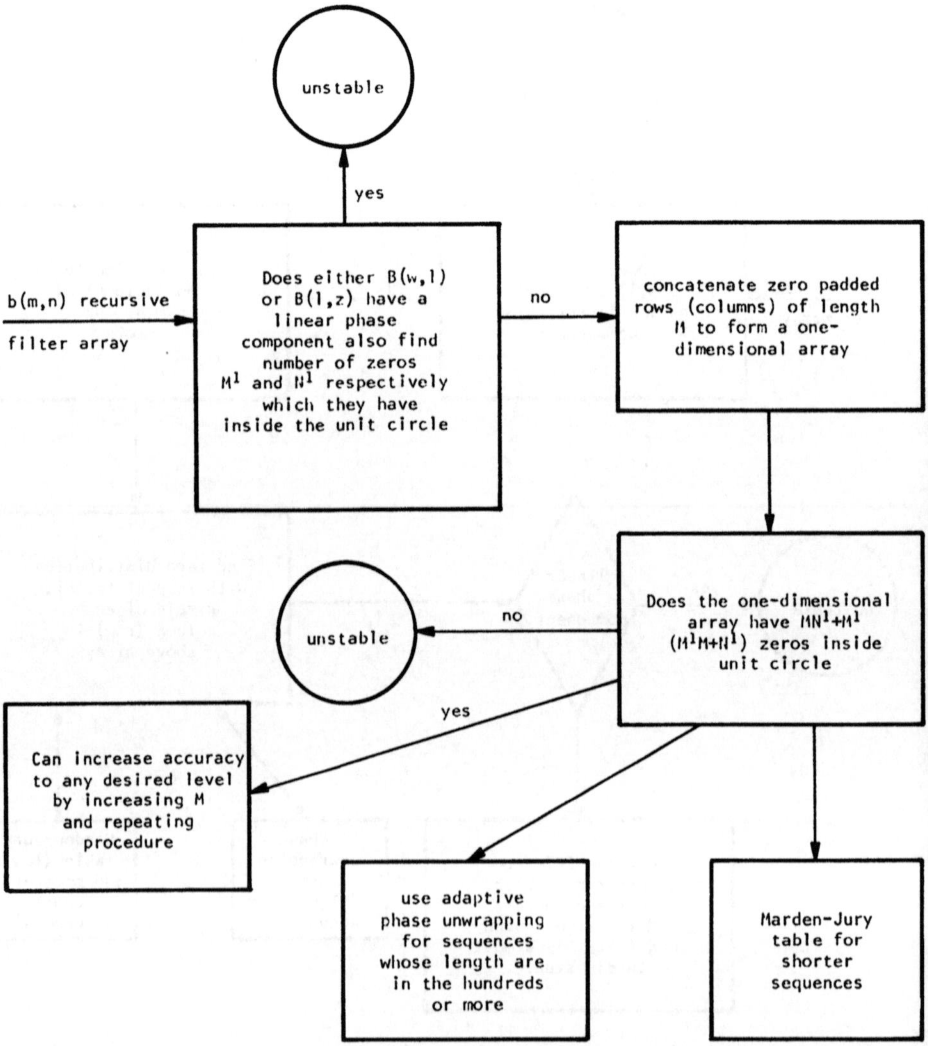

Fig. 4.15. Row and column concatenation algorithms

fail. Moreover, for filters of order greater than four or so in either dimension, the algebraic methods are very difficult to program and computational time becomes excessive. Hence, we believe that for moderate and high-order filters that these approximate or practical tests are the only feasible choice. Of all the practical tests, the above algorithms, in particular, those based on (4.51, 52) appear to be the most efficient. Figures 4.14 and 15 summarize these algorithms.

Below several examples are given to illustrate our algorithms.

Example 10 (E10)

$$B_1(w,z)=0.5w^{-1}z+1+0.85w+0.1wz+0.5w^2z^{-1}\,. \qquad \text{(E10.1)}$$

Condition 7(b):

$$B_1(w,1)=0.5w^{-1}+1+0.95w+0.5w^2$$

$$B_1(1,z)=0.5z^{-1}+1.85+0.6z\,.$$

Both have no linear phase components.

Condition 7(a): [row test]

Each row was transformed via a 1024 length FFT and the root distribution of the resulting columns revealed that, at least for this accuracy, $B_1(w,z)\neq 0$ on T^2 [since $b(m,n)$ is real only 513 root distribution tests are needed.] Therefore, $b(m,n)$ is "most likely" a stable filter array. A similar result is obtained if the column algorithm is used.

Condition 7(a): [concatenation tests]

(i) $B_1(\lambda,\lambda^M)=0.5\lambda^{2-M}+1+0.85\lambda+0.5\lambda^{M-1}+0.1\lambda^{M+1}$

(ii) $B_1(\lambda^N,\lambda)=0.5\lambda^{1-N}+1+0.85\lambda^N+0.1\lambda^{N+1}+0.5\lambda^{2N-1}\,.$

For these to have no linear phase terms the number of roots inside the unit circle for (i) should equal $M-2$ and $N-1$ for (ii) where M and N are the lengths of the zero padded rows and columns, respectively. Table 4.3 lists results of using the concatenation algorithm for different values of M and N. The root distributions were determined via a modified adaptive phase unwrapping

Table 4.3. Concatenation stability test on $b_1(m,n)$

M or N value	Length of row concatenated sequence	Length of column concatenated sequence	Ideal zero distribution inside unit circle row-column	M and N sequence zero distribution row-column
12	24	35	10 – 11	10 – 11
18	36	39	16 – 17	16 – 17
38	72	107	35 – 35	34 – 35
72	144	215	70 – 71	70 – 71
144	288	431	142 – 143	142 – 143
288	576	863	286 – 287	286 – —

algorithm [4.41, 16]. This algorithm efficiently implements the Nyquist test. Our experience [4.48] has been that this algorithm gives accurate root distribution information (also phase information) for sequences as long as six hundred on a PDP-11-45 with floating point mantissa of 24 bits. However, for longer sequences a larger floating point word size should be used to minimize FFT noise problems. Notice that the zero distribution always correspond to that of a stable filter.

Example 11 (E11)

$$B_2(w,z)=0.5w^{-1}z+1+0.89w+0.1wz+0.5w^2z^{-1}\,. \tag{E11.1}$$

$B_2(w,1)$ and $B_2(1,z)$ have no linear phase term, so condition 7(b) is satisfied. When the row algorithm is employed a phase discontinuity occurs for FFT sizes of 32 or larger. Therefore, $b_2(m,n)$ is definitely unstable. Alternately, this fact can be determined by using the column algorithm.

Condition 7(a): [concatenation tests]

Concatenations of zero padded rows and columns yield

(i) $B_2(\lambda,\lambda^M)=0.5\lambda^{2-M}+1+0.89\lambda+0.5\lambda^{M-1}+0.1\lambda^{M+1}$,

(ii) $B_2(\lambda^N,\lambda)=0.5\lambda^{1-N}+1+0.89\lambda^N+0.1\lambda^{N+1}+0.5\lambda^{2N-1}$.

If $b(m,n)$ is stable then the number of zeros inside the unit circle of the above polynomials should equal $M-2$ and $N-1$, respectively, for all values of M and N. Table 4.4 indicates results for different values of M and N. A phase discontinuity appears when $M \geqq 12$ and $N \geqq 18$ and hence, this test also indicates that there exists $(w_0,z_0)\in T^2$ such that $B(w_0,z_0)=0$, so $b(m,n)$ is unstable.

Table 4.4. Concatenation stability tests on $b_2(m,n)$

M or N value	Length of row concatenated sequence	Length of column concatenated sequence	Ideal zero distribution inside unit circle row-column	M and N sequence zero distribution row-column
6	12	17	4 – 5	4 – 7
12	24	35	10 – 11	8 – 11
18	36	53	16 – 17	14 – 19
36	72	107	34 – 35	32 – 41
72	144	215	70 – 71	68 – 81
144	288	431	142 – 143	138 – 165

Example 12 (E12)

$$B_3(w,z)=1-1.2w+0.5w^2-1.5z+1.8wz-0.75w^2z+0.6z^2-0.2718w^2z^2\,. \tag{E12.1}$$

The conditions of Theorem 7(b) are satisfied. The row algorithm detects a phase discontinuity for FFT sizes of 128 or larger while the column algorithm detects a discontinuity for FFT sizes of 32 or larger. The concatenation algorithms indicate a phase discontinuity for $B_3(\lambda,\lambda^M)$ for $M=18$ and for $B_3(\lambda^N,\lambda)$ for $N=36$. Therefore, all these algorithms indicate that $b(m,n)$ is unstable. It is interesting to note that $b_3(m,n)$ can be made stable by changing $b_3(2,2)$ by 0.0001 to 0.2719.

If the cepstrum test [4.8] is applied to this example no conclusion can be obtained for FFT sizes as large as 64×64. Hence the above algorithms are superior both in computation time and sensitivity.

It should be stressed that, in practice, it is only necessary to apply the row, column, row concatenation, or column catenation algorithms once for some suitable large FFT size L or values of M or N. This choice is dependent on the accuracy desired by the user. Indeed, there is a tradeoff between the accuracy of the test and computation time.

4.13 Finding the Zeros of the Two-Dimensional Fourier Transform

Several of the methods presented in the last section can be employed as efficient techniques for finding the zeros of $B(u,v)$ for $0<u<2\pi$ and $0\leqq v<2\pi$. Here, we will allow $b(m,n)$ to be any finite extent complex array. First apply (4.51) to locate discontinuity in zero distribution in z. This will specify intervals U_i on the u axis where zeros lie. Next apply (4.52) to locate discontinuities in zero distribution in w this specifying intervals V_i on the v axis. The widths of the intervals U_i and V_i depend on the FFT sizes used. Now, define $W_{ij}=U_i\times W_j$. At least half of the rectangles W_{ij}, will contain a zero of $B(u,v)$. See Fig. 4.16 for an example where $B(w,z)$ has two zeros on T^2. Finally, each W_{ij} can be searched for zeros to any desired accuracy using a minimization algorithm on $|B(u,v)|$ such as those discussed in [4.49]. Some standard methods are Fibonacci search, golden section search, pattern and one-at-a-time search, Powell's method, simplex methods, steepest decent algorithm, generalized Newton-Raphson, Fletcher-Powell algorithm, etc.

At this point the reader may ask, why not apply a minimization algorithm to the real function $|B(u,v)|$ directly instead? The answer is that standard minimization methods suffer from the following problems:

1) sensitivity to initial conditions,
2) tendency to converge to local minimum,
3) difficulties in finding all zeros of $B(u,v)$,
4) sensitivity of convergence to deviations from quadratic form.

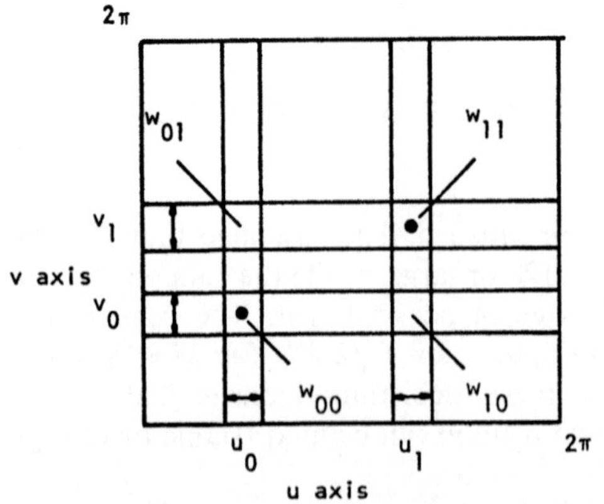

Fig. 4.16. Calculation of zeros of $B(u, v)$

Furthermore, since standard methods generally do not utilize the FFT algorithm then each new point must be computed directly from DFT. However, our procedure does not have problems with initial conditions and with finding all the zeros of $B(u, v)$. Furthermore, since we have restricted the area of search to be $2\pi/L$ by $2\pi/L$ where L is the FFT size in both the u and the v direction, problems with local minima and with convergence should be minimized. Moreover, this procedure will indicate whether $B(u, v)$ ever takes on a zero value, if not then no minimization algorithm need be applied. However, if a direct procedure is applied to $|B(u, v)|$ the conclusion that $|B(u, v)|$ does not have a zero can be made only after an extensive search of each local minimum and many choices of initial conditions.

There are basically two limiting cases where this algorithm fails. As mentioned earlier our algorithm doesn't work well when the zero set of $B(w, z)$ "barely" intersects the distinguished boundary T^2. The other case occurs when two zero sheets cross T^2 at approximately the same point in such a way that a phase discontinuity cannot be detected. However, if the FFT size is picked large enough these cases will rarely ever occur.

This chapter will be concluded with a section discussing the frequency response of general recursive filters.

4.14 Frequency Response of General Recursive Filters

The frequency response of a stable recursive filter defined by (4.7b) is

$$H(u, v) = \frac{A(u, v)}{B(u, v)},$$

where w is replaced by e^{-ju} and z is replaced by e^{-jv}. If $b(m, n)$ is a stable recursive filter array then $B(u, v)$ must satisfy certain stringent conditions. In particular, the gain of the filter averaged over certain directions in the frequency plane, must have no variation in a perpendicular direction [4.43]. These conditions were first reported by *Murray* [4.43]. In this paper, he studied the frequency response of first-quadrant filters. Below, we are going to extend and modify his development so that is applies to general recursive filters.

For our purposes Murray's theorem can be stated as follows:

Theorem 9. If $b(m,n)$ is a stable first-quadrant recursive filter array then

$$\frac{1}{2\pi}\int_0^{2\pi} \log|B(\mathrm{e}^{\mathrm{j}mu}, \mathrm{e}^{\mathrm{j}(nu+\psi)})|du = \log|b(0,0)|$$

for all ψ $(0 \leqq \psi < 2\pi)$ and for all integers $m>0$ and $n>0$.

His proof for this theorem relies heavily on the mean-value property of harmonic functions (i.e., the value of a harmonic function at the center of a circle is the average of the boundary values on the circle). We generalize this theorem to the following:

Theorem 10. Let $b(m,n)$ be a recursive filter array with $\beta^* = S[(M_1, N_1), (M_2, N_2)]$. Define $D = M_1N_2 - M_2N_1$, then

$$\frac{1}{2\pi}\int_0^{2\pi} \log|B(\mathrm{e}^{\mathrm{j}Pu}, \mathrm{e}^{\mathrm{j}(Qu+\psi)})|du = \log|b(0,0)|$$

for all ψ $(0 \leqq \psi \leqq 2\pi)$ and any given P and Q where $P = \mathrm{sgn}(D)N_2l_1 - \mathrm{sgn}(D)N_1l_2$; $Q = -\mathrm{sgn}(D)M_2l_1 + \mathrm{sgn}(D)M_1l_2$ with $l_i \in Z^+$ and $D \neq 0$ if $D=0$ $P = M_1l_1$, $Q = N_1l_2$.

Proof. By [4.16, 22] $R(\lambda) = B(\lambda^P, \lambda^Q)$ is a polynomial with $R(0) = b(0,0)$ and is nonzero for $|\lambda| \leqq 1$. Furthermore, since $G(w,z) \neq 0$ on $\bar{U}^2$ then the polynomial $B(\lambda^P, \lambda^Q \mathrm{e}^{\mathrm{j}\psi})$ also is nonzero for $|\lambda| \leqq 1$. Therefore, by Jensen's formula it follows that

$$\frac{1}{2\pi}\int_0^{2\pi} \log|B(\mathrm{e}^{\mathrm{j}Pu}, \mathrm{e}^{\mathrm{j}(Qu+\psi)})|du = \log|b(0,0)| \,. \quad \text{QED}$$

This theorem can be reinterpreted as follows: since any projection of a stable recursive filter array $b(m,n)$, which is a one-dimensional causal sequence, must be minimum phase by Theorem 4.6 then by Projection Slice Theorem [4.47] and Jensen's formula the corresponding slice in the Fourier plane must satisfy the above equation. Let $P_{(M,N)}(\lambda) = \sum\sum b(m,n)\lambda^{Mm+Nn}$ be called the (M,N) projection of $b(m,n)$ where M and N are mutually prime. If $b(m,n)$ is a recursive filter array with $\beta^* = S[M_1, N_1), (M_2, N_2)]$ then define $\hat{S}[(M_1, N_1),$

$(M_2, N_2)]$ as

$$\hat{S}[(M_1, N_1), (M_2, N_2)]$$
$$= S[(-\operatorname{sgn}(D)N_1, \operatorname{sgn}(D)M_1), (\operatorname{sgn}(D)N_2, -\operatorname{sgn}(D)M_2)]$$

if $D \neq 0$ and

$$= C(r, s) \in Z^2 : rM_1 + sN_1 \geqq 0\}$$

if $D = 0$.

It readily follows that $\hat{S}[(M_1, N_1), (M_2, N_2)]$ holds all indices (M, N) for which $P_{(M,N)}(\lambda)$ is a causal projection. Therefore, from Theorem 4.10 if

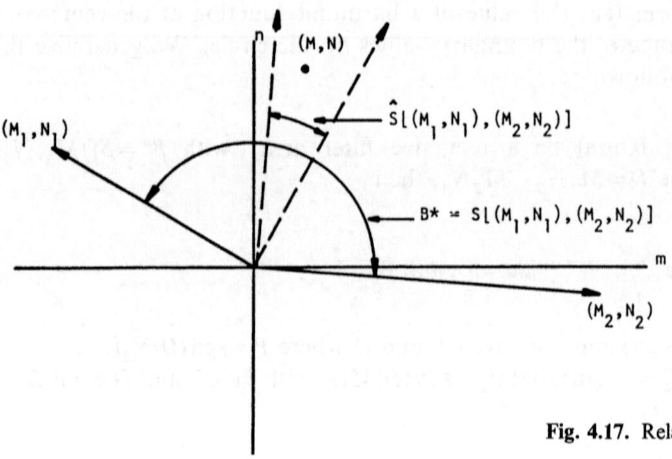

Fig. 4.17. Relation between S and $\hat{S}$

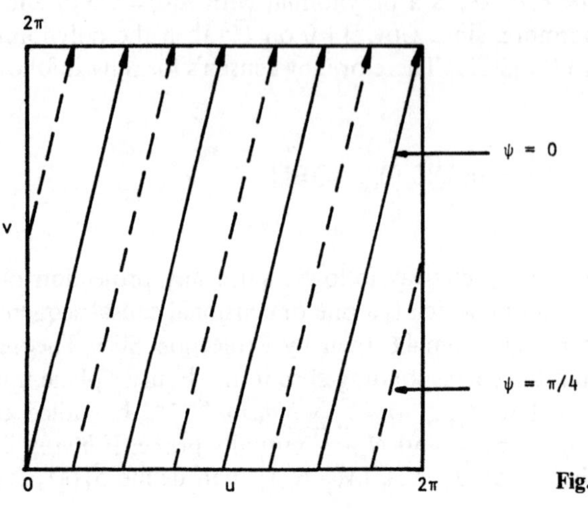

Fig. 4.18. Illustration of Theorem 10

$(M,N) \in \hat{S}$ then the logarithmic gain of the Fourier transform along the direction (M,N) is zero. Furthermore, the larger the angle of β^* is the narrower $\hat{S}$ is so $b(m,n)$ can approximate more general frequency responses. Figure 4.17 exemplifies the relation between S and $\hat{S}$, while Fig. 4.18 illustrates for a typical $(M,N) \in \hat{S}$ over which values in the frequency the average logarithmic magnitude gain is zero [assuming $b(0,0)=1$] for $\psi=0$ and $\psi=\pi/4$.

Appendix A: One-Dimensional Polynomial Root Distribution Tests

In this appendix several z plane methods for determining the zero distribution with respect to the unit circle will be reviewed. In particular, emphasis will be placed on the Schur–Cohn criterion and the Jury table test. Schur developed two such stability tests [4.34, 50]. In the first, positive definiteness of a nth order Hermitian form defined using the coefficients of the polynomial $F(z)$ is checked. For the second test a specific sign pattern of n determinants of order $2, 4, \ldots, 2n$ is checked. Cohn generalized Schur's results by noting that the rank and signature of Schur's Hermitian form and the sign pattern of Schur's determinants give information about the zero distribution of a polynomial with respect to the unit circle. Later, Fujiwara rederived and extend Schur–Cohn's results in a simple manner [4.50]. *Jury* [4.33] simplified Schur–Cohn's determinant method for real coefficients by approximately halving the order of the determinants. Moreover, he formulated this technique as an inners matrix [4.51, 52]. More recently, *Anderson* and *Jury* [4.50] simplified Schur–Cohn's Hermitian matrix formulation for real coefficients by halving the size of matrix to be checked for positive definiteness.

Below, the Schur–Cohn Hermitian criterion will be considered because there exists an important extension of it to two dimensions. Associated with $F(z)$ is the $n \times n$ Hermitian matrix $C=(\gamma_{ij})$ defined by

$$\gamma_{ij} = \sum_{p=1}^{i} (a_{n-i+p} a^*_{n-j+p} - a^*_{i-p} a_{j-p}) \quad i \leqq j\,, \tag{A.1}$$

where a_i is the coefficient of the z^i term of $F(z)$. Then the number of zeros z_i of $F(z)$ for which $|z_i|<1$ and for which z_i^{-1} is not also a zero is the number of positive eigenvalues of C; the number of zeros z_i for which $|z_i|>1$ and for which z_i^{-1} is not also a zero is the number of negative eigenvalues of C; the number of zeros z_i for which either $|z_i|=1$ or z_i^{-1} is also a zero is the nullity of C [4.13]. Therefore, $F(z)$ has no zeros inside the unit circle if and only if C is negative definite so the leading principal minors must alternate their signs. Hence, the Schur–Cohn criterion requires either calculating the eigenvalues or computing approximately n determinants of orders up to n.

In all of the above root distribution tests, a number of determinants of order as high as $n/2$ (sometimes $2n$) must be computed. Generally, even the most efficient determinant algorithms require $O(n^3)$ multiplications and additions for an nth order matrix. In fact Gauss elimination requires approximately $n(n^2-1)/3$ calculations [4.53]. For higher order polynomials these methods are very inefficient. Their only redeeming value rests on their ability to consisely present a closed form stability criterion.

There do exist several efficient algebraic zero distribution tests. A division method [4.33] was developed by *Jury* and provides an approach where stability and root distribution information can be determined by examining the results of an iterative division process. Another method, similar to the division algorithm, was formulated by *Marden* and modified by *Jury* [4.33, 34]. This procedure requires the construction of a table (called Marden–Jury table or just Jury table) by the calculation of many two by two determinants. The sign variations in the first column of this table give the root distribution information. For an nth order polynomial this procedure only requires $(n+1)\cdot n$ multiplications. Furthermore, fewer calculations may result if only stability information is desired. The table test consists of the following:

$$\begin{array}{cccccc}
z^0 & z^1 & z^2 & \dots & z^{n-1} & z^n \\
a_0 & a_1 & a_2 & & a_{n-1} & a_n \\
a_n^* & a_{n-1}^* & a_{n-2}^* & & a_1^* & a_0^* \\
b_0 & b_1 & b_2 & & b_{n-1} & \\
b_{n-1}^* & b_{n-2}^* & b_{n-3}^* & & b_0^* & \\
 & \vdots & & & & \\
r_0 & r_1 & & & & \\
r_1^* & r_0^* & & & & \\
t_0 & & & & &
\end{array} \tag{A2}$$

where $b_k = \begin{vmatrix} a_0 & a_{n-k} \\ a_n^* & a_k^* \end{vmatrix}$, etc.

Let $P_1 = \mathrm{sgn}(b_0)$, $P_2 = \mathrm{sgn}(b_0 \cdot c_0), \dots, P_n = \mathrm{sgn}(b_0 \cdot c_0 \dots t_0)$ then the number of zeros of $F(z)$ inside the unit circle is equal to the number of negative P_j provided all P_j are nonzero. If any $P_j = 0$ then the table test can be modified to give the polynomial's root distribution; however, this procedure is more complicated [4.33]. A necessary and sufficient condition for $F(z)$ to have all its zeros inside the unit circle is that $b_0 < 0$, $c_0 > 0, \dots, t_0 > 0$ so the procedure may terminate early.

This table test is far more efficient than the determinant tests and appears to be one of the most efficient algebraic tests available for determining the root distribution of a polynomial with respect to the unit circle. Approximately N^2 multiplications and $N^2/2$ additions are needing to determine a $N$the order polynomial's root distribution. Again depending on whether the coefficients are real or complex these operations should be interpreted correspondingly.

References

4.1 D. Dudgeon: "Two-dimensional Recursive Filtering"; Sc.D. Thesis, M.I.T., E.E. Dep., Cambridge, MA, May (1974)
4.2 R. Mersereau, D. Dudgeon: Proc. IEEE **63**, 610–623 (1975)
4.3 T. Huang: IEEE Trans. Au-**20**, 158–163 (1972)
4.4 J. Shanks, S. Treitel, J. Justice: IEEE Trans. AU-**20**, 115–128 (1972)
4.5 J. Justice, J. Shanks: IEEE Trans. AC-**18**, 284–286 (1973)
4.6 R. DeCarlo, J. Murray, R. Saeks: Int. J. Control **25**, 657–675 (1977)
4.7 G. Maria, M. Fahmy: IEEE Trans. ASSP-**21**, 470–472 (1973)
4.8 M. Ekstrom, J. Woods: IEEE Trans. AASP-**24**, 115–128 (1976)
4.9 N. Bose: Proc. IEEE **65**, 824–840 (1977)
4.10 D. Goodman: IEEE Trans. CAS-**24**, 201–208 (1976)
4.11 J. Costa, A. Venetsanopoulous: IEEE Trans. ASSP-**24**, 424–425 (1976)
4.12 D. Dudgeon: IEEE Trans. ASSP-**25**, 476–484 (1977)
4.13 B. Anderson, E. Jury: IEEE Trans. AU-**21**, 366–372 (1973)
4.14 D. Goodman: IEEE Trans. ASSP-**24**, 426–427 (1976)
4.15 M. Strintzis: IEEE Trans. CAS-**24**, 432–437 (1977)
4.16 B.T. O'Connor: "Techniques for Determining the Stability of Two-Dimensional Recursive Filters and Their Application to Image Restoration"; Ph.D. Thesis, Purdue University, E.E. Dept., West Lafayette, IN, May (1978)
4.17 D. Davis: IEEE Trans. ASSP-**24**, 425–426 (1976)
4.18 R. DeCarlo: "An Algebraic Topological Approach to Stability Theory"; Ph.D. Thesis, E.E. Department, Texas Tech., Lubbock, Texas (1976)
4.19 R. DeCarlo, R. Saeks, J. Murray: Proc. IEEE **65**, 978–979 (1977)
4.20 E. Jury, O. Chen: IEEE Trans. CT-**20**, 476–480 (1973)
4.21 R. Churchill: *Complex Variables and Applications* (McGraw-Hill, New York 1974)
4.22 W. Rudin: *Function Theory in Polydiscs* (Benjamin, New York 1969)
4.23 D. Dudgeon; IEEE Trans. ASSP-**23**, 242–243 (1975)
4.24 T. Huang, W. Schreiber, O. Tretiak: Proc. IEEE **59**, 1586–1609 (1971)
4.25 M. Strentzis: "Conditions for Stability of Nonanticipative Two-Dimensional Recursive Filters", IEEE Trans. Aut. Control. (to be published)
4.26 D. Siljak: IEEE Trans. ASSP-**23**, 473–474 (1974)
4.27 D. Siljak: IEEE Trans. CAS-**22**, 185–189 (1975)
4.28 N. Bose: IEEE Trans. ASSP-**25**, 117–120 (1977)
4.29 H. Ansell: IEEE Trans. CT-**11**, 214–223 (1964)
4.30 D. Goodman: Proc. IEEE **66**, 796–797 (1978)
4.31 E. Guillemin: *Synthesis of Passive Networks* (Wiley, New York 1957)
4.32 B. O'Connor, T. Huang: IEEE Trans. ASSP-**26**, 380–381 (1978)
4.33 E. Jury: *Theory and Application of the z-transform Method* (Wiley, New York 1964)
4.34 M. Marden: *The Geometry of the Zeros of a Polynomial in a Complex Variable* (American Math. Soc., New York 1949)
4.35 E. Horowitz, S. Sahni: J. Assoc. Comput. Mach. **22**, 38–50 (1975)
4.36 H. Schussler: IEEE Trans. ASSP-**24**, 87–89 (1976)
4.37 A. Ralston, H. Wilf: *Mathematical Methods for Digital Computers* (Wiley, New York 1967)
4.38 R. Read, S. Treitel: IEEE Trans. GE-**11**, 153–160 (1973)
4.39 M. Lal, H. Singh, R. Panwar: IEEE Trans. CAS-**22**, 62–63 (1975)
4.40 A. Bliss: *Algebra Functions* (Amer. Math. Society, New York 1933)
4.41 J. Tribolet: IEEE Trans. ASSP-**25**, 170–177 (1977)
4.42 A. Oppenheim, R. Schafer: *Digital Signal Processing* (Prentice-Hall, Englewood Cliffs, N.J. 1975)
4.43 J. Murray: "Spectral Factorization and Quarter-Plane Digital Filters", Tech. Rpt', Texas Tech., Lubbock, Texas (1977)

4.44 B.O'Connor, T.Huang: "Two-dimensional Complex Cepstrum", in Image Understanding and Information Extraction, Quart. Prog. Rpt. Purdue University, May–July (1976) F 30602-75-C-0150, pp. 103–105
4.45 P.Pistor: IBM J. Res. Dev., pp. 59–71, January (1974)
4.46 M.Ahmadi, R.King: Proc. IEEE **65**, 893–898 (1977)
4.47 R.Mersereau, A.Oppenheim: Proc. IEEE, **62**, 1319–1338 (1974)
4.48 R.Mersereau, D.Dudgeon: IEEE Trans. ASSP-**22**, 320–325 (1974)
4.49 J.Bandler: IEEE Trans. MIT-**17**, 533–552 (1969)
4.50 B.Anderson, E.Jury: IEEE Trans. AC-**18**, 157–163 (1973)
4.51 E.Jury: Proc. IEEE **63**, 1044–1068 (1975)
4.52 E.Jury: *Inners and Stability of Dynamic Systems* (Wiley, New York 1975)
4.53 B.Anderson, E.Jury: IEEE Trans. AU-**21**, 366–372 (1973)
4.54 B.O'Connor, T.Huang: IEEE Trans. ASSP-**26**, 550–560 (1978)

5. Two-Dimensional Kalman Filtering

J. W. Woods

With 31 Figures

In the recent past there has been much interest in the problem of two-dimensional (2-D) recursive filtering with particular application to images. Much of this work has centered on the extension of Kalman filtering to two dimensions. Early attempts [5.1–3] to achieve a truly recursive 2-D Kalman filter were of only limited success due to both the difficulty in establishing a suitable 2-D model and also the high dimension of the resulting state vector. In fact a straightforward extension of the one-dimensional (1-D) Kalman filter techniques results in a number of state variables proportional to N for the filtering of an $N \times N$ digital image.

In [5.4], two new algorithms were presented which to a large extent overcame the computational problems that had preculded the use of 2-D Kalman-like processors. Both vector and scalar scanning methods were considered, but emphasis was placed on the scalar scan because it leads to processors which are recursive in both dimensions, i.e., 2-D recursive processors. These results have application to a wide range of problems involving estimation on both space and space–time data fields, including the image processing problem where most emphasis is placed in this chapter. The Kalman algorithms allow the use of space-variant models which provide a better match to local source statistics leading to a greater noise reduction.

Other approaches to recursive image estimation include the work of *Strintzis* [5.5], *Jain* [5.6], and *Murphy* and *Silverman* [5.7]. In [5.5], *Strintzis* presented an autoregressive-moving average (ARMA) modeling approach to recursive processing coupled with an interesting additive rather than multiplicative based spectral decomposition. The addition of a moving average to the AR model of Kalman filtering provides some modeling advantages. However the developed model is for the correlation function, not the process itself. Thus simulation is not possible. Also only the steady state is considered, where the image processing problem can be embedded in the class of cyclic processes. Unfortunately this embedding cannot be done if nonhomogeneous image models are considered or if random boundary conditions are explicitly treated.

In [5.6], *Jain* considered both implicit and explicit partial difference equation models for images. The models are compared and evaluated for estimation improvement on representative images. The computational methods are based on fast transform algorithms. The method is not suitable for spatially varying or nonhomogeneous image models and it is restricted to deterministic boundary conditions.

In [5.7], *Murphy* and *Silverman* considered a vector scanning approach to image restoration. Two approximations are presented. A strip filter is developed which is similar to the one developed in [5.4]. Also a constrained or reduced update vector processor is developed which complements nicely the scalar reduced update processor developed in [5.4]. Also [5.7] treats restoration while [5.4] does not explicitly consider the restoration problem. However [5.13] extends the 2-D reduced update filter to image restoration.

5.1 Overview

In this chapter the 2-D Kalman based algorithms of [5.4] are presented in some detail. Also a new multiple model algorithm is presented for space-variant images which can offer quite dramatic numerical and subjective improvement.

In Sect. 5.2 the relevant Markov image models are presented for both space-invariant and -variant, Gaussian and non-Gaussian cases. A new doubly stochastic model consisting of a lower level 2-D Markov chain and upper level multiple models is presented.

Section 5.3 treats spatial Kalman filtering. It starts with a brief review of the concept of state and its role in 1-D Kalman filtering. Then the 2-D Kalman scalar and vector filters are defined and their undesirable computational properties are pointed out. Next the Kalman strip filter and the reduced update Kalman filter are derived and some examples are presented.

The reduced update filter is then singled out for further investigation. This filter is extended to the inverse type problem of image restoration. The uniquely 2-D boundary value problems are discussed. The section closes with extensions to space-variant filtering by employing multiple models using both an adaptive and a doubly stochastic approach.

Section 5.4, System Identification, treats the problem of estimating the model parameters for the various recursive estimation approaches mentioned above. The first identification method is spectral modeling using 2-D spectral factorization methods to design the recursive models. Next linear predictive methods are presented for spatial domain approximation. The section ends with a discussion of new methods for identification of multiple models.

Section 6.5 summarizes the results of this chapter and makes suggestions for future research directions.

5.2 Markov Image Models

This section is divided into two parts. The first part is on space-invariant, Gaussian models and the second part is on space-variant models, both Gaussian and non-Gaussian.

5.2.1 Space-Invariant, Gaussian Models

Here we present a summary of the properties of discrete Markov random fields. It will be shown that the nonsymmetric half-plane (NSHP) recursive models generate a special class of the Markov random fields presented in [5.8]. The first-order and separable models, so often used in image processing, are thus seen to be special cases of the general discrete Markov random field.

Definition. A discrete Markov random field: Let s be a random field on I^2, the 2-D lattice. Let a band of minimum width p, ∂G ("the present") separate I^2 into two regions G^+ ("the future") and G^- ("the past"). Then s is Markov-p if $\{s|_{G^+}$ given $s|_{\partial G}\}$ is independent of $\{s|_{G^-}\}$ for all ∂G (see Fig. 5.1).

In [5.8], it is shown that for the homogeneous, zero-mean, Gaussian case, this definition is equivalent to the following interpolative difference equation model,

$$s(m,n)=\sum_{D_p} c_{kl}s(m-k,n-l)+u(m,n)\,, \tag{5.1}$$

where

i) $\boldsymbol{E}[s(m,n)\,u(k,l)]=\sigma_u^2\delta_{mk}\delta_{nl}$,

ii) $D_p=\{k,l|k^2+l^2\leqq p^2$ and $(k,l)\neq(0,0)\}$,

iii) $u(m,n)$ is a Gaussian, zero-mean, homogeneous random field with correlation function of bounded support,

$$R_u(m,n)=\begin{cases}\sigma_u^2, & (m,n)=(0,0),\\ -c_{mn}\sigma_u^2, & (m,n)\in D_p,\\ 0, & \text{elsewhere}.\end{cases}$$

iv) The c_{kl} are the interpolator coefficients of the minimum mean square error (mmse) linear interpolation problem:[1] Given $\{s$ for $(m,n)\neq(0,0)\}$, estimate $s(0,0)$.

The solution is given as the conditional mean,

$$\boldsymbol{E}[s(0,0)|s \;\text{ for }\; (m,n)\neq(0,0)]=\sum_{D_p} c_{kl}s(-k,-l)\,.$$

In this formulation σ_u^2 is the mean square interpolation error.

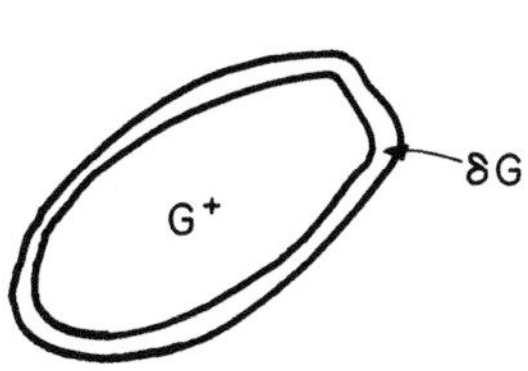

Fig. 5.1. Example of regions used in definition of Markov random field

1 Note that the field is assumed to be Gauss Markov-p. Also the linear equations for the interpolator will involve correlation values over D_{2p}.

Using the above properties it was shown [5.8] that

$$R_s(m,n) = \sum_{D_p} c_{kl} R_s(m-k, n-l) + \delta_{mn}\sigma_u^2 \tag{5.2}$$

so that upon taking the Fourier transform of (5.2) we arrive at

$$S_s(u,v) = \sigma_u^2 \Big/ \left\{1 - \sum_{D_p} c_{kl} \exp[j(ku+lv)]\right\} \tag{5.3}$$

which is the spectral density function of a Markov-p random field.

The spectrum of (5.3) is not generally factorable into finite-order NSHP factors [5.9]. Thus to every 2-D Markov field, there does not correspond a finite-order causal (NSHP sense) model. However, the inverse argument is trivially true. A finite-order NSHP model driven by spatially white noise generates a Markov random field. The implication of this result for modeling seems clear. Instead of fitting (5.3) to given spectral data, one should instead take a factored form and then fit this to the data. This then converts the modeling problem to one of filter design which can be accomplished, using a magnitude square criterion, by the methods of [5.10]. We also note that the generality of NSHP models assures us that for any smooth spectrum bounded away from $+\infty$ and 0, we can obtain a convergent sequence of NSHP models using 2-D spectral factorization, followed by windowing and truncation to ever-increasing NSHP support. Thus we arrive at general Markov image models of the form

$$s(m,n) = \sum_{\mathscr{R}_{\oplus +}} c_{k,l}\, s(m-k, n-l) + w(m,n)\,, \tag{5.4}$$

where $\mathscr{R}_{\oplus +} = \{m \geqq 0, n \geqq 0\} \cup \{m < 0, n > 0\}$ as shown in Fig. 5.2 and where $w(m,n)$ is a white noise field. Thus the compact state region of Fig. 5.1 has expanded by factorization into the very large global state region of Fig. 5.2. This is to be expected since the state region must separate the *past* from the *future* however these two regions are configured. We note that when the factored Markov model is of finite order, then only a finite subset of the state is used at each iteration of (5.4). This region will be called the *local state region.* This region then denotes the support of the right-hand side of (5.4) for a (M_1, M_2)th order $\oplus +$ model of the NSHP variety.

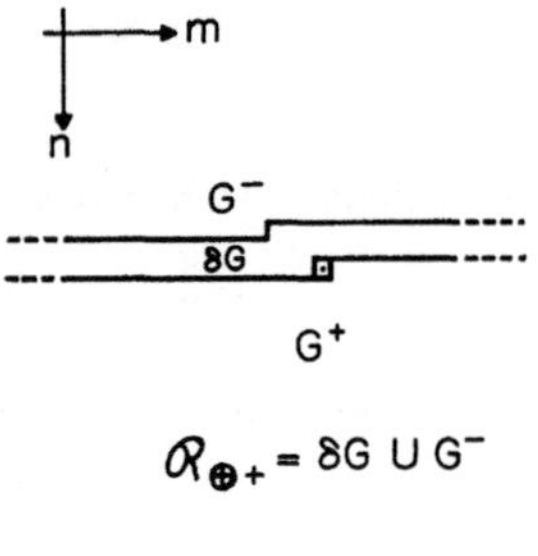

Fig. 5.2. Global state region of Markov nonsymmetric half plane (NSHP) model

Computationally the local and global states require two different types of memory. The local state must be stored in random access memory with size $= 2M_1M_2$ words. The global state however may be stored in lower cost shift register type memory using M_2 shift registers, each one a line or N words long.

In Sect. 5.4, methods will be presented for obtaining these models based on L^2 approximation in either the frequency domain or the space domain.

5.2.2 Space-Variant Models

In this section various space-variant models will be developed for 2-D recursive estimation. The corresponding identification problems will be discussed in Sects. 5.4.2 and 3.

Gaussian Models

The above Markov models can be made space variant by allowing the model coefficients $\{c_{kl}\}$ to be a function of position. The model would thus be denoted $\{c_{kl}^{mn}\}$. Then the 2-D Kalman recursive estimation methods to be presented in Sect. 5.3 could be directly applied to solve signal-in-noise problems with the models. In order to use such space-variant models, however, several problems must be overcome. The first problem is to obtain the model coefficients. This problem can be solved by the techniques of system identification. A second serious problem is the storage requirements for the model coefficients. For example, if the model was of 2×2th order, thus having 13 coefficients, the amount of storage required for the coefficients would be equivalent of 13 $N\times N$ digital pictures (or more depending on the word length required to store the coefficients) assuming that the coefficients changed at each pixel. This second problem dictates that the first problem, that of coefficient estimation, must be performed on-line from the noisy data as a part of the 2-D recursive estimation thereby alleviating the coefficient storage problem, see Fig. 5.3. The resulting processor would then properly be called a nonlinear processor, but to the extent that the model coefficients could be reliably determined by the coefficient estimator, then the resulting processor can be thought of as an approximation to the space-variant linear processor that is appropriate for space-variant Gaussian models.

The coefficient estimation problem is difficult because the noisy observations introduce a bias into the coefficient estimates obtained using linear prediction methods. Some system identification as well as some recursive estimation results are presented in Sect. 5.4.2.

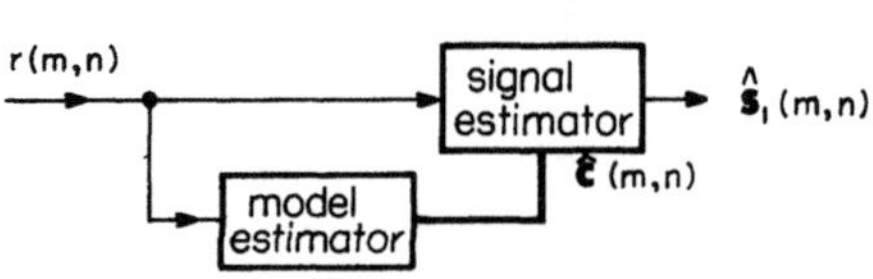

Fig. 5.3. Space-variant filtering with on-line model coefficient estimation

Spatial Markov Chain

For cases in which the spatial variation of the model parameters is either random or too rapid to estimate reliably, it becomes advantageous to just use a fixed small number of models. This in fact is the basis for the multiple-model extension of the 2-D Kalman filter to be discussed in Sect. 5.3.5.

To complete this model it is necessary to have a description for the underlying process which determines the elementary model to be used at each pixel. Such a description is the 2-D Markov chain. This is a generalization of the familiar 1-D Markov chain to two-dimensions. It generates a discrete valued random field $l(m, n)$ which specifies the one of say L models to be used at pixel (m, n).

The simplest example of sufficient generality would be a 2-D Markov chain with local state as shown in Fig. 5.4. Such a local state would require an $L^4 \times L$ stochastic matrix to specify its transition probabilities. This is because the (1×1)th order local state in Fig. 5.4 is four dimensional, thus giving a total of L^4 possible local states. Of these L^4 only L are possible new local states at the next pixel $(m+1, n)$. For the experimental cases to follow, $L=5$ was found to work rather well. In this case the transition matrix is 625×5. Each row of the transition matrix indicates the conditional probability of going from each of the L^4 possible present local states at (m, n) to the conditionally possible local states at $(m+1, n)$. The 2-D Markov chain is thus specified by giving the support of the local state and its transition matrix $\boldsymbol{P}$.

The 2-D Markov chain can be used to model a partitioned data field where regions of various shapes and orientations provide a large scale structure to an image. Within these regions, the various Gaussian submodels can then be used to provide a fine structure or texture to the image. Another possiblity is to use most models to model the edge regions of images with one or more submodels left over to model the nonedge or interior regions. This composite model could have the advantage of allowing optimal smoothing without blurring the edges in images. The filtering for such compound models can usefully be thought of as a two-step process: at a given pixel one first estimates the local state of the underlying Markov chain, then one chooses a most likely model to do the

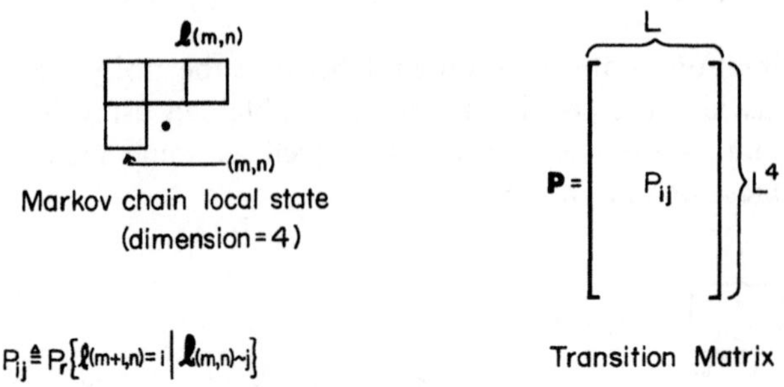

Fig. 5.4. Diagram illustrating 2-D Markov chain of order 1×1

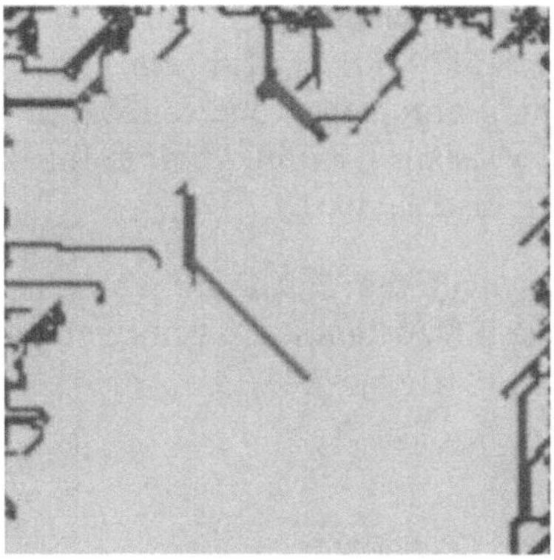

Fig. 5.5. Simulation of Markov chain model for edges of geometric image

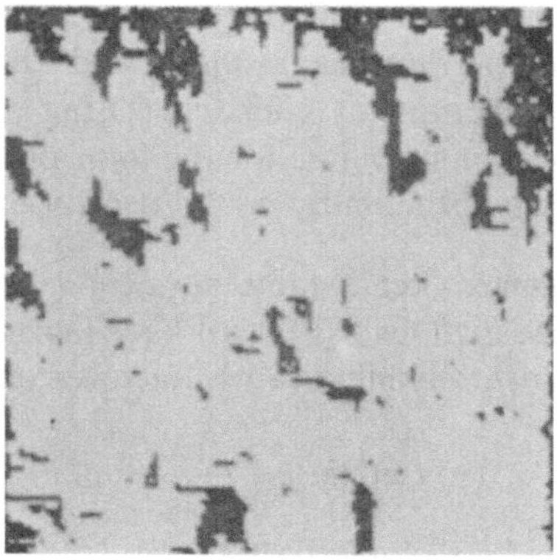

Fig. 5.6. Simulation of Markov chain model for edges of face image

estimation of the higher level random field. This is the application made in Sect. 5.3.5 where the Markov chain is used in a multiple-model extension of the reduced update Kalman filter.

Here we present some examples of 2-D Markov chain simulations which were conducted to initially explore this approach. There are two examples generated from a geometrical and a natural image shown in Fig. 5.21a and Fig. 5.22a, respectively. Figures 5.5 and 6 show Markov chain data, whose transition matrices were computed from the images of Fig. 5.21a and 5.22a, respectively, to model the edge structure. The value of L was taken equal to 5, with the first four l values indicating edges at 0°, 45°, 90°, and 135°, respectively. The value $l=5$ was used to indicate nonedge or isotropic regions. The steady state only holds in the interior of the images as the boundary conditions, on the top and sides, were chosen to be jointly independent.

Doubly Stochastic Gaussian Field

Using the 2-D Markov chain to specify an underlying structure, one may generate a doubly stochastic model by using L different sets of parameters $\{c_{ij}^{l}\}$ to generate the random signal field. The resulting signal model would become

$$s(m,n)=\sum_{\mathscr{R}_{\oplus+}} c_{ij}^{l(m,n)} s(m-i,n-j)+u(m,n)\;, \tag{5.5}$$

where $l(m,n)$ is the 2-D Markov chain field with local state support consisting of a finite extent NSHP region similar to the local state support of (5.4). Thus given $\{l(m,n)\}$, the resulting signal model is space-variant Gaussian. The composite model consisting of the Markov chain and (5.5) is doubly stochastic and is termed *doubly stochastic Gaussian* in analogy with the terminology in the doubly stochastic Poisson case.

Thus the 2-D doubly stochastic Gaussian model consists of a lower level 2-D Markov chain $\{l(m,n)\}$ and a higher level 2-D Markov field $\{s(m,n)\}$ which is conditionally Gaussian. Using conditioning arguments, one can derive the following formula for the joint probability $p[\{s(m,n), l(m,n)\}]$. (Note this is a mixed probability with a density part and a discrete part.)

Lemma. Let a finite segment $[0,M]\times[0,N]$ of the 2-D doubly stochastic Gaussian field $\{(s(m,n)\}$ have lower-level chain field $\{l(m,n)\}$. Then the overall joint probability can be factored as follows:

$$p[\{s(m,n), l(m,n)\}]$$

$$\prod_{m,n}(P[l(m,n)|\boldsymbol{l}_1(m-1,n)]p[s(m,n)|\boldsymbol{s}_1(m-1,n), l(m,n)]) ,$$

where $P[\cdot|\cdot]$ is the transition probability assignment for the Markov chain as determined by the elements of its transition matrix $\boldsymbol{P}$ and $p[\cdot|\cdot,l]$ is the $(1,0)$–step prediction density for the lth submodel, $\{c^l_{ij}\}$, for $1\leqq l\leqq L$. Also $\boldsymbol{s}_1$ and $\boldsymbol{l}_1$ are, respectively, the local states of the conditionally Gaussian field and the Markov chain.

Proof. By the conditioning expansion we can write

$$\begin{aligned} p[\{s(m,n), l(m,n)\}] &= p[s(0,0), l(0,0), s(1,0), l(1,0), \ldots, s(M,N), l(M,N)] \\ &= p[s(0,0), l(0,0)]\, p[s(1,0), l(1,0)|s(0,0), l(0,0)] \\ &\quad \cdots p[s(m,n), l(m,n)|s \text{ and } l \text{ on past}] \\ &\quad \cdots p[s(M,N), l(M,N)|s \text{ and } l \text{ on past}] \\ &= \prod_{m,n} p[s(m,n), l(m,n)|s \text{ and } l \text{ on past}] . \end{aligned}$$

Now

$$\begin{aligned} p[s(m,n), l(m,n)|s \text{ and } l \text{ on past}] &= P[l(m,n)|s \text{ and } l \text{ on past}] \\ &\quad \cdot p[s(m,n)|l(m,n) \text{ and } s \text{ and } l \text{ on past}] \end{aligned}$$

by definition of conditional probability. Next

$$P[l(m,n)|s \text{ and } l \text{ on past}] = P[l(m,n)|\boldsymbol{l}_1(m-1,n)]$$

by definition of underlying 2-D Markov chain and

$$p[s(m,n)|l(m,n) \text{ and } (s \text{ and } l \text{ on past})] = p[s(m,n)|l(m,n), \boldsymbol{s}_1(m-1,n)]$$

by definition of the 2-D doubly stochastic Gaussian field.

Therefore

$$p[s(m,n), l(m,n)|s \text{ and } l \text{ on past}]$$
$$= P[l(m,n)|\boldsymbol{l}_1(m-1,n)]\,p[s(m,n)|s_1(m-1,n), l(m,n)]$$

and the decomposition follows. This decomposition proves useful in solutions of the system identification problems for multiple models to be presented in Sect. 5.4.3.

5.3 Kalman Filters

Kalman filtering is well established for the one-dimensional case. The key assumption is that the signal can be modeled by a first-order vector dynamical system. The vector is called the state vector and summarizes the effect of the past on the future for this 1-D Markov process. This dynamical model is given as [5.11]:

$$\boldsymbol{s}(m) = \boldsymbol{C}\boldsymbol{s}(m-1) + \boldsymbol{w}(m)\,, \tag{5.6}$$

$$\boldsymbol{r}(m) = \boldsymbol{H}\boldsymbol{s}(m) + \boldsymbol{v}(m)\,, \tag{5.7}$$

where $\boldsymbol{s}$ is the signal state vector, $\boldsymbol{r}$ is the observation or received signal, $\boldsymbol{v}$ is the observation noise, and $\boldsymbol{w}$ is the random process which generates the signal. The matrices $\boldsymbol{C}$, and $\boldsymbol{H}$ are, respectively, the system and observation matrices.

In (5.6, 7) $\boldsymbol{w}$ and $\boldsymbol{v}$ are uncorrelated, white Gaussian, zero-mean noise sources with correlation

$$\boldsymbol{E}[\boldsymbol{w}(m)\boldsymbol{w}^{\mathrm{T}}(k)] = \boldsymbol{Q}_w\delta_{mk} \quad \text{and} \quad \boldsymbol{E}[\boldsymbol{v}(m)\boldsymbol{v}^{\mathrm{T}}(k)] = \boldsymbol{Q}_v\delta_{mk}\,,$$

where $\boldsymbol{Q}_w$ and $\boldsymbol{Q}_v$ are correlation matrices and δ_{mk} is the Kronecker delta function. The Kalman filtering equations for this model are [5.11]:

$$\hat{\boldsymbol{s}}_b(m) = \boldsymbol{C}\hat{\boldsymbol{s}}_a(m-1)\,, \tag{5.8}$$

$$\boldsymbol{P}_b(m) = \boldsymbol{C}\boldsymbol{P}_a(m-1)\boldsymbol{C}^{\mathrm{T}} + \boldsymbol{Q}_w\,, \tag{5.9}$$

$$\boldsymbol{K}(m) = \boldsymbol{P}_b(m)\boldsymbol{H}^{\mathrm{T}}(\boldsymbol{H}\boldsymbol{P}_b(m)\boldsymbol{H}^{\mathrm{T}} + \boldsymbol{Q}_v)^{-1}\,, \tag{5.10}$$

$$\hat{\boldsymbol{s}}_a(m) = \hat{\boldsymbol{s}}_b(m) + \boldsymbol{K}(m)[\boldsymbol{r}(m) - \boldsymbol{H}\hat{\boldsymbol{s}}_b(m)]\,, \tag{5.11}$$

$$\boldsymbol{P}_a(m) = [\boldsymbol{I} - \boldsymbol{K}(m)\boldsymbol{H}]\boldsymbol{P}_b(m)\,, \tag{5.12}$$

where

$$\boldsymbol{P}_i(m) \triangleq \boldsymbol{E}[(\boldsymbol{s}-\hat{\boldsymbol{s}})_i(\boldsymbol{s}-\hat{\boldsymbol{s}})_i^{\mathrm{T}}]\,, \qquad i = a, b$$

is the error covariance matrix. The subscripts a and b indicate *after* and *before* updating, respectively. Note these equations assume $\boldsymbol{s}$ is zero mean. If the mean were not zero, (5.8–12) are correct with the understanding that $\hat{\boldsymbol{s}}$ and $\boldsymbol{r}$ are centered at their means.

These equations have the following simple interpretation. First in (5.8) we project the last estimate forward using the dynamics of the system model. Then in (5.11) we update this estimate using the new observation $\boldsymbol{r}(m)$. Equation (5.10) gives the gain matrix for the updating in (5.11). The remaining equations (5.9) and (5.12) are the error covariances necessary to calculate the new gain matrix.

5.3.1 Kalman Filtering In Two Dimensions

A Scalar 2-D Kalman Filter [5.4]

When we try to extend the above formalism to two dimensions we encounter a problem concerning the 1-D definition of state as we will now show. First we consider a particular scanning of a discrete 2-D square region consisting of an $N \times N$ regularly spaced lattice. Since the scanning does not qualitatively affect the results, we will assume the raster scan, i.e., left to right, advance one line, then repeat. Continuing with this procedure for the next $N-2$ lines we complete the scan. Thus at any point in the picture some points will be the "past", one point will be the "present", and the remaining points will be the "future". These words will thus have their conventional meaning with respect to the order in which the points are processed, and will correspond to the Markov field regions presented in Fig. 5.2. Next we generalize our dynamical model consisting of (5.6), using the 2-D NSHP $(M \times M)$th order model of (5.4) which is to be compared to the scalar version of (5.6),

$$s(m) = \sum_{k=1}^{M} c_k s(m-k) + w(m) . \tag{5.13}$$

Now we are in a position to appreciate the difficulty with state in two dimensions. For while (5.4) only uses $O(M^2)$ points for each computation of a new output point, so that the amount of computation is related to the order of the model as in (5.13), when (5.4) moves across the image, values of s will be needed from the previous M lines. Thus the memory requirements of (5.4) are not of order related to the order of the model as in (5.13). Because of this large memory requirement, the state for (5.4) must contain the M previous lines and hence has $O(MN)$ components. It is convenient to call this full state the *global state* and to take the part of the state on the support of the model and call it the *local state*.

Local State: The minimum part of the global state needed to compute the next output given the present input.

This local state has a number of elements related to the order of the model and hence its size determines the order of computation of the model. However, the storage requirements are related to the size of the full or global state. In the ordinary 1-D case the local state vector is the global state.

Next we will consider the Kalman filtering equations for this model. We will find that both the per point computation and the memory grow with N^2, the square of the width of the image field. Because of this unsatisfactory situation, we then turn to some approximate filters which greatly reduce the computation while only slightly affecting the accuracy of the estimate.

To obtain the Kalman filtering equations for the signal model of (5.4) and the observation equation

$$r(m,n) = s(m,n) + v(m,n) \ , \tag{5.14}$$

where w and v are independent, white, Gaussian sources, we note that the scanning operation transforms the 2-D problem into an equivalent 1-D problem. Thus (5.4) can be put into the form of (5.6) by defining the global state vector of $M(N+1)$ components

$$\begin{aligned} \boldsymbol{s}(m,n) = [&s(m,n), s(m-1,n), \ldots, s(1,n); \quad s(N,n-1), \ldots, s(1,n-1); \ldots; \\ &s(N,n-M), \ldots, s(m-M+1, n-M)]^{\mathrm{T}} \ . \end{aligned} \tag{5.15}$$

With this transformation we obtain

$$\boldsymbol{s}(m,n) = \boldsymbol{C}\boldsymbol{s}(m-1,n) + w(m,n) \ , \tag{5.16}$$

with corresponding observational model

$$r(m,n) = \boldsymbol{h}^{\mathrm{T}}\boldsymbol{s}(m,n) + v(m,n) \tag{5.17}$$

which is identical in form to (5.6, 7) with the line by line scanning understood. Thus, we could immediately write down the Kalman equations. They would be (5.8–12) with the above interpretation of the $\boldsymbol{s}$ vector. The difficulty with these equations is the amount of computation and memory requirements associated with them. For example consider an $(M \times M)$th order system model as in (5.4) and an observation region consisting of an $N \times N$ square with $N \gg M$. Then the dimension of the matrix equations (5.8–12) is approximately MN. This means that in general the order of the computation is $O(M^3N^3)$. However, taking advantage of the spatial invariant structure of the signal model this is reduced to $O(M^2M^2)$ as will be shown below. For $M=4$ and $N=100$, we require on the order of 160,000 multiplies and adds per output point. The overall total computation for the 10,000 element picture would be $O(10^9)$. At 1 μs/operation, the computer time for such a calculation would be 10^3 s or approximately 20 min. In addition, the data storage problems are immense. To store $\boldsymbol{P}_{b,a}$ at each stage we need $O(MN)^2$ storage locations. For the above M and N, this is

160,000 words of storage to be accessed at each picture point! These numbers tell us that the exact 2-D Kalman scalar filter is computationally unmanageable, given today's state of the computer art. But not only that, it is also wasteful of computation as will be pointed out below.

Next we will consider the 2-D Kalman filter for vector observations. This is also an exact processor as is the scalar one mentioned above, however, the vector processor observes a line at a time instead of a point at a time.

A Kalman Filter for Vector Observations

In this formulation we consider a vector scanning [5.12] consisting of a (horizontal) line at a time, as is necessary to do optimal Kalman vector filtering. We can write a recursive signal model

$$\boldsymbol{s}(n) = \boldsymbol{C}\boldsymbol{s}(n-1) + \boldsymbol{w}(n) \tag{5.18}$$

with vector observations

$$\boldsymbol{r}(n) = \boldsymbol{H}\boldsymbol{s}(n) + \boldsymbol{v}(n)\,. \tag{5.19}$$

Here

$$\begin{aligned}\boldsymbol{s}(n) \triangleq [&s(1,n), \dots, s(N,n);\ s(1,n-1), \dots, s(N,n-1); \dots;\\ &s(1,n-M+1), \dots, s(N,n-M+1)]^{\mathrm{T}}\end{aligned}$$

and

$$\boldsymbol{r}(n) = [r(1,n), \dots, r(N,n)]^{\mathrm{T}}\,.$$

Since (5.18, 19) are of the form of (5.6, 7), the Kalman equations are (5.8–12) with the above interpretations for the $\boldsymbol{s}$ and $\boldsymbol{r}$ vectors. We note that these matrix equations are of order $O(MN)$ as in the 2-D Kalman scalar filter. However, each interation of the vector filter yields N estimates (a whole line), so for the example given for the scalar filter, the vector filter computation times are reduced by a factor of N yielding an $O(M^3N^2)$ computation. The storage requirements are the same and hence this is a computationally unmanageable solution[2]. We will see below how splitting the picture up into partially overlapping strips can greatly reduce the computation of this vector filter. Further reductions in the order of the computation can be realized by using the special block form of the matrices $\boldsymbol{C}$ and $\boldsymbol{H}$ in (5.18, 19). These reductions in computation result in allowing the practical implementation of the Kalman vector filter. As in the 2-D scaler observation case, we can construct the recursive dynamical model (5.18) by using 2-D spectral factorization.

---

2 Note that the term vector processor was used in [5.12] to refer to a processor less than a line wide. However we will reserve this term for the full width processor.

5.3.2 Approximations

In this section are summarized approximate methods from [5.4] for realizing the 2-D Kalman scalar and vector filters. While both these techniques provide 2 to 3 orders of magnitude reduction in computation and storage, experience has indicated that the reduction in the optimality of the estimate is small. The reason for this surprising result is that, as we will see, the cause of the excessive computations is "purely theoretical", that is the very uncommon, but possible, occurrence of high correlation at great distances. An approximate Kalman vector filter called the strip filter is first developed. Then a scalar filter, called the reduced update Kalman filter, is presented.

Strip Filter

The Kalman strip filter arises from the following considerations. Firstly, although the vector filter has a state vector the width of the picture, which is required for theoretical optimality, experience suggests that very few images will have the correlation at great distances to justify such computation. Rather, most images show significant correlation only over rather compact neighborhoods of a given point. Thus we consider processing strips of the picture independently. The strips are overlapped with only the middle points being used as the final estimate. If the strip width is greater than the "correlation distance" of the signal and noise field, then close to optimum performance can result for points near the center. Processing the picture in these strips can greatly reduce the order of the computation and storage.

Secondly, we observe that if the data are stationary in the horizontal direction we can save computation and storage by the following scheme. Take the strips in the vertical direction and note that the filters for (5.9, 12) and the gain equation (5.10) need only be calculated for the first strip. Then the corresponding line in each of the remaining strips may be updated via (5.11) using the same gain matrix. In this way storage need only be provided for the covariance and gain matrices in the first strip.

Thus, these two simplifications allow us to construct a good approximation to the Kalman vector filter which is computationally manageable. Next the details of constructing such a Kalman strip filter will be presented. This will consist of two parts: first, a discussion of the model; second, a discussion of the filter for the model.

The Strip Dynamical Model

The recursive signal model is the same as given in (5.18) with vector observations given in (5.19) with the exception that the vectors now are of dimensions MW, where W is the strip width ($W \ll N$). The model development starts from a general NSHP recursive model and develops from it a vector recursive model,

$$\boldsymbol{s}(n)=\boldsymbol{C}\boldsymbol{s}(n-1)+\boldsymbol{G}\boldsymbol{w}(n)\,, \tag{5.20}$$

$$\boldsymbol{r}(n)=\boldsymbol{H}\boldsymbol{s}(n)+\boldsymbol{v}(n)\,, \tag{5.21}$$

where

$$\boldsymbol{s}(n)\triangleq[s(1,n),\ldots,s(W,n);\ s(1,n-1),\ldots,s(W,n-1);\ldots;\\ s(1,n-M+1),\ldots,s(W,n-M+1)]^{\mathrm{T}}$$

and

$$\boldsymbol{r}(n)=[r(1,n),\ldots,r(W,n)]^{\mathrm{T}}\,.$$

To obtain $r=s+v$ choose

$$\boldsymbol{H}=[I:0:\ldots:0]^{\mathrm{T}}\,.$$

Now $\boldsymbol{C}$, $\boldsymbol{G}$, and $\boldsymbol{Q}_w$ must be chosen so that s approximately satisfies the equation for the $(M\times M)$th order NSHP signal model is given by (5.4).

The conversion from (5.4) to (5.20) is most easily seen in the Z-transform domain. Taking the Z transform of (5.4) we see that the system function transforming s into w is

$$W(z_1,z_2)/S(z_1,z_2)=1-C(z_1,z_2)\,, \tag{5.22}$$

where

$$C(z_1,z_2)=\sum_{\mathcal{R}_{\oplus+}} c_{pq}z_1^{-p}z_2^{-q}\,.$$

Equation (5.22) can be factored into

$$1-C(z_1,z_2)=[1-C(z_1,\infty)]\,[1-C^{(1)}(z_1,z_2)]\,, \tag{5.23}$$

where

$$C(z_1,\infty)=\sum_{p>0} c_{p0}z_1^{-p}$$

and

$$C^{(1)}(z_1,z_2)\ \text{contains only}\ q>0\ \text{terms}\,.$$

Now $C^{(1)}$ can be directly represented by $\boldsymbol{C}$ since it represents an operation on the previous rows. The first factor can be inverted and represented by $\boldsymbol{G}$. In this way one can derive the following form for $\boldsymbol{C}$,

$$\boldsymbol{C}=\begin{bmatrix} \boldsymbol{A}_1 & \boldsymbol{A}_2 & \dots & \dots & \boldsymbol{A}_M \\ \boldsymbol{I} & \boldsymbol{0} & \dots & \dots & \boldsymbol{0} \\ \boldsymbol{0} & \boldsymbol{I} & \boldsymbol{0} & & \boldsymbol{0} \\ \boldsymbol{0} & \dots & \boldsymbol{0} & \boldsymbol{I} & \boldsymbol{0} \end{bmatrix} \tag{5.24}$$

with $W \times W$ blocks $\boldsymbol{I}$ and $\boldsymbol{A}_M$; $m=1,\dots,M$. Here

$$\boldsymbol{A}_m=[\boldsymbol{I}-\boldsymbol{C}_0]^{-1}\boldsymbol{C}_m\ , \tag{5.25}$$

where the $\boldsymbol{C}_m$ are given by

$$(\boldsymbol{C}_m)_{ij}=c_{m,i-j}\ . \qquad (5.26$$

In like manner we could set

$$\boldsymbol{G}=[(\boldsymbol{I}-\boldsymbol{C}_0)^{-1}:\boldsymbol{0}:\boldsymbol{0}:\dots:\boldsymbol{0}]^{\mathrm{T}}\ .$$

However, the edge effects due to the small strip width W would be expected to cause significant errors. To lessen this problem we choose $\boldsymbol{G}$ to make the correlations of (5.4) and (5.20) agree exactly in the current row. This procedure leaves uncorrected slight errors in the correlations with the $(M-1)$ previous rows in the model. Since the errors are both small and concentrated near the strip edges, they are expected to be insignificant in effect on the "saved" estimates in the middle section of the strip. An alternative modelling approach is the method of *Murphy* and *Silverman* [5.7] based on full half-plane models.

The above results on the Kalman strip filter can be summarized in the following theorem.

Theorem 1. The Kalman equations for the strip model are

$$\hat{\boldsymbol{s}}_b^i(n)=\boldsymbol{C}\hat{\boldsymbol{s}}_a^i(n-1)\ , \tag{5.27}$$

$$\hat{\boldsymbol{s}}_a^i(n)=\hat{\boldsymbol{s}}_b^i(n)+\boldsymbol{K}^1(n)[\boldsymbol{r}^i(n)-\boldsymbol{H}\hat{\boldsymbol{s}}_b^i(n)] \tag{5.28}$$

$$\boldsymbol{P}_b^i(n)=\boldsymbol{C}\boldsymbol{P}_a^i(n-1)\boldsymbol{C}^{\mathrm{T}}+\boldsymbol{G}\boldsymbol{Q}_w\boldsymbol{G}^{\mathrm{T}}\ , \tag{5.29}$$

$$\boldsymbol{K}^i(n)=\boldsymbol{P}_b^i(n)H^{\mathrm{T}}[\boldsymbol{H}\boldsymbol{P}_b^i(n)\boldsymbol{H}^{\mathrm{T}}+\boldsymbol{Q}_v]^{-1}\ , \tag{5.30}$$

$$\boldsymbol{P}_a^i(n)=[\boldsymbol{I}-\boldsymbol{K}^i(n)\boldsymbol{H}]\boldsymbol{P}_b^i(n)\ , \tag{5.31}$$

where i is the number of the strip $L=1,\dots,[N/M]$.

We note that for the horizontally stationary case, (5.29–31) need only be computed once for each line. While the estimates (5.27, 28) are computed for all strips at the nth line before proceeding to the $(n+1)st$ line.

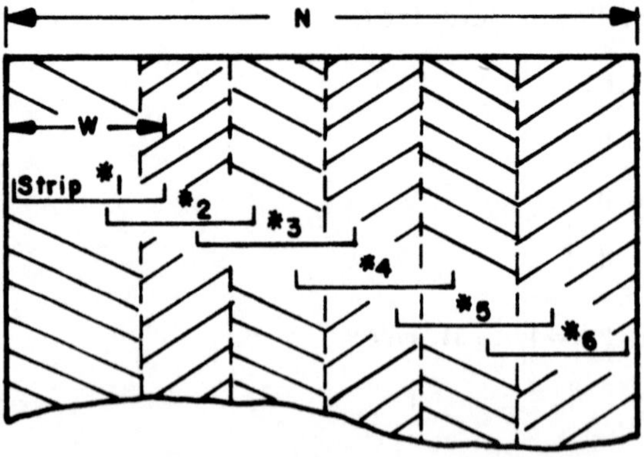

Fig. 5.7. Diagram for strip processing showing overlap of strips

Initial conditions are needed for (5.27, 29) at the first line. They are given by

$$s_b^i(1) = E[s^i(1)] = \mathbf{0}$$

and

$$\boldsymbol{P}_b^1(1) = \boldsymbol{E}[\boldsymbol{s}^1(1)\boldsymbol{s}^{1\mathrm{T}}(1)] = \boldsymbol{R}_s\,,$$

where $\boldsymbol{R}_s$ is the $MW \times MW$ correlation matrix of the model.

Figure 5.7 shows a diagram for strip processing where the middle third of the elements are retained for the final estimate of s. The overlap would increase computation by a factor of 3. In the setup used for an example to be presented below, a width of $W=25$ is used with retention of the middle 15 elements. In this case the "extra" computation due to the overlap is reduced to a factor of 2. The amount of "extra" computation can be traded off for uniformity in the estimation error.

Optimization of Strip Width

To minimize the amount of computation for the completely nonstationary case, we can choose the number of retained elements N_R and the strip width W consistent with obtaining near to optimum performance. To keep the saved points at least D from the strip edge, where D is the "correlation distance" of the signal s, we get

$$W = 2D + N_\mathrm{R}\,.$$

The number of computations per point is proportional to

$$(2D + N_\mathrm{R})^3/N_\mathrm{R} \quad \text{for} \quad N \gg 2D\,.$$

The minimum occurs at $N_\mathrm{R}^0 = D$ which yields

$$W^0 = 3D\,. \tag{5.32}$$

When the data is stationary horizontally then the number of computations can be asymptotically

$$(2D+N_{\mathrm{R}})^2/N_{\mathrm{R}} \quad \text{for} \quad N \gg 2D\,.$$

The minimum turns out to be $N_{\mathrm{R}}^0=2D$ which yields

$$W^0=4D\,. \tag{5.33}$$

Example

A noisy picture was processed with the Kalman strip filter. The signal was a 100×100 element center segment from a standard SMPTE test picture. The 8 bit original had an estimate of its mean removed prior to processing. The picture was then scaled so that its variance equaled 1. White Gaussian noise was added at unit variance to produce a SNR = 0 db. The original with mean

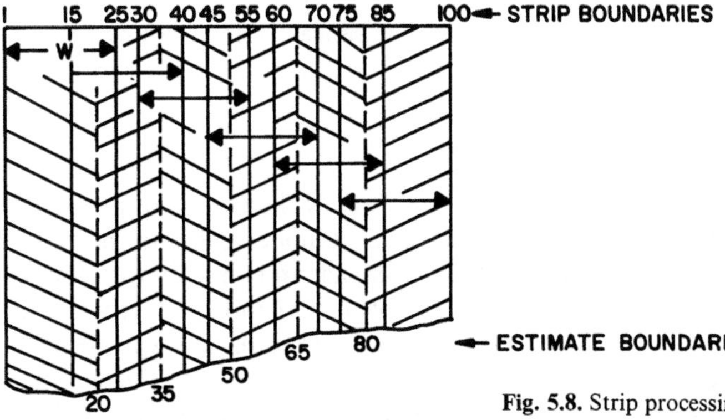

Fig. 5.8. Strip processing diagram for example

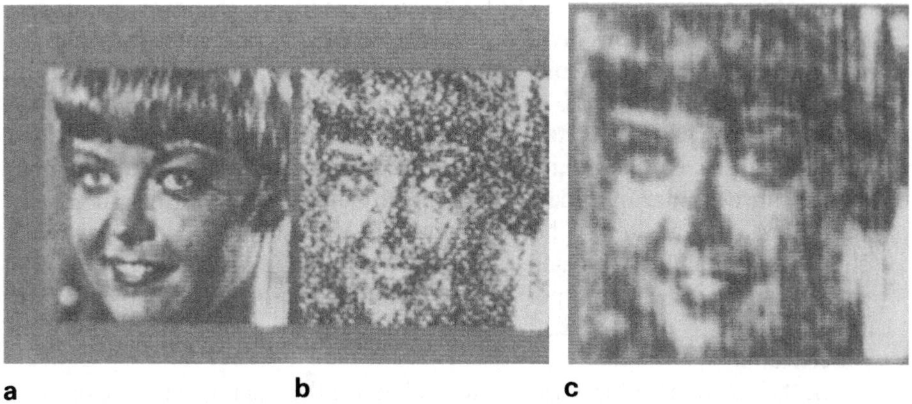

Fig. 5.9a–c. Strip processing example: **(a)** original image; **(b)** noisy image (Odb); **(c)** strip estimate

removed was also used to generate a (10×4)th order $+\oplus$ NHSP model. This turned out to be a model for the transposed pictures. Thus the noisy image was transposed and inserted into the Kalman strip filter program with the above NSHP model. The strip arrangement is shown in detail in Fig. 5.8. There are 6 strips, each 25 points wide, with 5 element overlap on each side for the interior strips. The middle 15 elements of the strips cover the picture and provide a fairly uniform error covariance.

The mean value was added and the estimate images transposed for presentation. Figure 5.9a shows the original image, Fig. 5.9b shows the noisy image and Fig. 5.9c shows the strip filter estimate. The measured SNR improvement was 7.7 db. As the model and noise were stationary, convergence was obtained, in this case in 10 lines. The computer time was under 15 s on a CDC 7600. The error covariance values were within 3% of their minimum value over the 15 element middle region of the strip. On the strip edge, it was as much as 60% above the minimum error variance value which was near the center, thus pointing out the advisability of overlapping the strips.
The 7.7 db improvement is about 1–2 db less than reported in [5.6] for a semicausal model. However the image used in [5.6] was the 64×64 center segment of the image used here, and this difference could account for the increased improvement.

Reduced Update Filter [5.4]

This section treats an approximation to the 2-D Kalman scalar processor presented in Sect. 5.3.1. This involves the dynamical model (5.4) and the scalar observation (5.14). As indicated earlier the global state for this filter is $O(MN)$ dimensional where M is the order of the recursive model and N is the width of the picture. Thus all these points must be updated in an equation equivalent to (5.11). The main concept of the reduced update filter derives from the fact that the Kalman equations are composed of two steps: a prediction part and an update part. Now the prediction part, as set forth in (5.8), is a computationally straigthforward projection of $O(M^2)$ previous estimates. However, the update part involves calculations involving each of the $O(MN)$ random variables in the global state. Since $N \gg M$, one can eliminate the bulk of the computation by reducing the update process. Thus one chooses to only update those elements of the global state within a certain distance of the point currently being processed (m, n). It can be expected that this procedure will result in a good approximation because significant updates will generally be confined to a region around the observation at (m, n). Therefore omitting the update of points far away should only minimally impact performance. For convenience of notation this region is taken to be the support of the local state at (m, n). Hence there are only $O(M^2)$ points to update, i.e., we only update the local state vector for each observation.

In the remainder of this section we will derive the equations for the optimal updating of the local state vector. We will see that these equations can point the

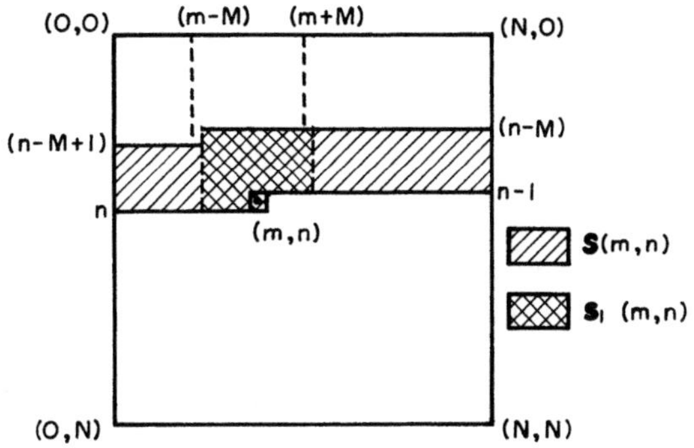

Fig. 5.10. Assignment of points to local s_1 and to global s state

way to further simplifications which have been found to result in only slightly suboptimum performance.

Let the signal model be $M \times M$th order NSHP as generated by (5.4). Let the field $\{r(m,n)\}$ in (5.14) be observed over an $N \times N$ rectangular region. Next introduce a vector notation for the local state vector corresponding to a scalar line-by-line scan.

$$\begin{aligned} s_1(m,n) \triangleq [&s(m,n), \ldots, s(m-M+1,n); \\ &s(m+M+1,n-1) \ldots s(m-M+1,n-1); \\ &\ldots: s(m+M+1,n-M), \ldots, s(m-M+1,n-M)]^{\mathrm{T}} . \end{aligned} \tag{5.34}$$

Order the remaining points of the global state onto $s_2(m,n)$. Thus the resulting assignment of points is as shown in Fig. 5.10.

With this convention the global state vector can be written

$$s(m,n) \triangleq [s_1^{\mathrm{T}}(m,n), s_2^{\mathrm{T}}(m,n)]^{\mathrm{T}} . \tag{5.35}$$

The state dynamical model can then be written as in (5.16), where $\boldsymbol{C}$ is a system propagation matrix determined by $\{c_{kl}\}$ and the ordering of the state vector $s(m,n)$. Note that (5.16) will hold for all the points (m,n) except near the boundaries where boundary conditions must be incorporated. The drive vector is given as

$$w(m,n) \triangleq [w(m,n), 0, \ldots, 0]^{\mathrm{T}} . \tag{5.36}$$

The scalar observation equation is (5.17) where $\boldsymbol{h}^{\mathrm{T}} = (1, 0, \ldots, 0)$. We can partition the matrix $\boldsymbol{C}$ similarly to s,

$$\boldsymbol{C} = \begin{bmatrix} \boldsymbol{C}_{11} & \boldsymbol{C}_{12} \\ \boldsymbol{C}_{21} & \boldsymbol{C}_{22} \end{bmatrix} . \tag{5.37}$$

It then turns out that C_{11} and C_{12} contain all the $\{c_{kl}\}$ terms and the remainder of C constitutes a shift transformation. Thus (5.16) may be rewritten as

$$s_1(m,n)=C_{11}s_1(m-1,n)+w_1(m,n)+C_{12}s_2(m-1,n)\;, \tag{5.38}$$

where w has been partitioned similarly to s. Equation (5.38) focuses on the computation to be performed at (m,n). That is, the new s_2 requires only the shifting of previously computed values.

We can similarly partition $h^{\mathrm{T}}=(h_1^{\mathrm{T}}, h_2^{\mathrm{T}})$ with $h_2=0$ to get a new observation equation

$$r(m,n)=h_1^{\mathrm{T}}s_1(m,n)+v(m,n)\;. \tag{5.39}$$

Derivation of Reduced Update Filter

Assume that the received array $\{r(m,n)\}$ is scanned in line-by-line fashion. Then the Kalman filtering equations can be written immediately from (5.8–12).

Extrapolation: $m \to m+1$

$$P_b(m,n)=CP_a(m-1,n)C^{\mathrm{T}}+Q_w\;, \tag{5.40}$$

$$\hat{s}_b(m,n)=C\hat{s}_a(m-1,n)\;. \tag{5.41}$$

Update:

$$k(m,n)=P_b(m,n)h[h^{\mathrm{T}}P_b(m,n)h+\sigma_v^2]^{-1}\;, \tag{5.42}$$

$$\hat{s}_a(m,n)=\hat{s}_b(m,n)+k(m,n)[r(m,n)-h^{\mathrm{T}}\hat{s}_b(m,n)]\;, \tag{5.43}$$

$$P_a(m)=[I-k(m,n)h^{\mathrm{T}}]P_b(m,n)\;. \tag{5.44}$$

If one uses an arbitrary gain matrix instead of the one prescribed in (5.42), the covariance P_a given in (5.44) changes to

$$P_a(m,n)=(I-k(m,n)h^{\mathrm{T}})P_b(m,n)(I-k(m,n)h^{\mathrm{T}})^{\mathrm{T}}+k(m,n)\sigma_v^2k^{\mathrm{T}}(m,n)\;. \tag{5.45}$$

Let $k(m,n)$ have the form

$$k(m,n)=\begin{bmatrix} k_1(m,n) \\ \hline 0 \end{bmatrix}\;, \tag{5.46}$$

where the partitioning corresponds to that of $s(m,n)$. We now choose $k_1(m,n)$ to minimize the trace of $P_a(m,n)$ as given in (5.45) for $k(m,n)$ as given in (5.46). The result can be summarized as the following theorem.

Theorem 2. The reduced update Kalman filter equations for the model of (5.4) and observations given by (5.39) are:

$$\boldsymbol{P}_b(m,n)=\boldsymbol{C}\boldsymbol{P}_a(m-1,n)\boldsymbol{C}^{\mathrm{T}}+\boldsymbol{Q}_w\,, \tag{5.47}$$

$$\hat{\boldsymbol{s}}_{1b}(m,n)=\boldsymbol{C}_{11}\hat{\boldsymbol{s}}_{1a}(m-1,n)+\boldsymbol{C}_{12}\hat{\boldsymbol{s}}_{2a}(m-1,n)\,. \tag{5.48}$$

Update:

$$\boldsymbol{k}_1(m,n)=\boldsymbol{P}_{11,b}(m,n)\boldsymbol{h}_1[\boldsymbol{h}_1^{\mathrm{T}}\boldsymbol{P}_{11,b}(m,n)\boldsymbol{h}_1+\sigma_v^2]^{-1}\,, \tag{5.49}$$

$$\hat{\boldsymbol{s}}_{1a}(m,n)=\hat{\boldsymbol{s}}_{1b}(m,n)+\boldsymbol{k}_1(m,n)[r(m,n)-\boldsymbol{h}_1^{\mathrm{T}}\hat{\boldsymbol{s}}_{1b}(m,n)]\,, \tag{5.50}$$

$$\boldsymbol{P}_{11,a}(m,n)=[\boldsymbol{I}-\boldsymbol{k}_1(m,n)\boldsymbol{h}_1^{\mathrm{T}}]\boldsymbol{P}_{11,b}(m,n)\,, \tag{5.51a}$$

$$\boldsymbol{P}_{12,a}(m,n)=[\boldsymbol{I}-\boldsymbol{k}_1(m,n)\boldsymbol{h}_1^{\mathrm{T}}]\boldsymbol{P}_{12,b}(m,n)\,, \tag{5.51b}$$

where $\boldsymbol{P}_a$ and $\boldsymbol{P}_b$ have been partitioned similarly to $\boldsymbol{s}$.

Scalar Equations

These equations (5.47–51) can provide great computational savings over the standard Kalman filtering equations. To understand these equations better, it is helpful to convert them back to scalar notation. First we note that (5.48) will become, in scalar notation,

$$\hat{s}_b^{(m,n)}(m,n)=\sum_{\mathscr{R}_{\oplus+}} c_{kl}\hat{s}_a^{(m-1,n)}(m-k,n-l)\,, \tag{5.52}$$

since it represents propagation of the previous estimates through the dynamics of the system. In these scalar equations, the superscript indicates the step in the filtering, while the argument represents the position of the data. Equation (5.47) represents the error in this predicted estimate, thus it becomes

$$R_b^{(m,n)}(m,n;k,l)=\sum_{op} c_{op}R_a^{(m-1,n)}(m-o,n-p;k,l)\,,\quad (k,l)\in\mathscr{S}_{\oplus+}^{(m,n)}\,, \tag{5.53}$$

$$R_b^{(m,n)}(m,n;m,n)=\sum_{kl} c_{kl}R_b^{(m,n)}(m,n;m-k,n-l)+\sigma_w^2\,, \tag{5.54}$$

where $\mathscr{S}_{\oplus+}^{(m,n)}$ is the support of the global state $\boldsymbol{s}(m,n)$. Equation (5.49) computes a $\boldsymbol{k}_1(m,n)$ which has the same support as $\boldsymbol{s}_1(m,n)$ namely, $\mathscr{R}_{\oplus+}^{(m,n)}$, the support of the local state vector. The scalar equation identical to (5.49) is

$$k^{(m,n)}(i,j)=R_b^{(m,n)}(m,n;m-i,n-j)/(R_b^{(m,n)}(m,n;m,n)+\sigma_v^2),\quad (i,j)\in\mathscr{R}_{\oplus+}^{(m,n)} \tag{5.55}$$

Similarly (5.50) becomes

$$\hat{s}_a^{(m,n)}(i,j)=s_b^{(m,n)}(i,j)+k^{(m,n)}(m-i,n-j)[r(m,n)-\hat{s}_b^{(m,n)}(m,n)]\,,$$
$$(i,j)\in\mathscr{R}_{\oplus+}^{(m,n)}\,. \tag{5.56}$$

Finally, (5.51a, b) both are expressible as the set of scalar equations

$$R_a^{(m,n)}(i,j;k.l)=R_b^{(m,n)}(i,j;k,l)-k^{(m,n)}(m-i,n-j)R_b^{(m,n)}(m,n;k,l)\,,$$
$$\text{for}\quad (i,j)\in\mathscr{R}_{\oplus+}^{(m,n)}\,;\ (k,l)\in\mathscr{S}_{\oplus+}^{(m,n)}\,. \tag{5.57}$$

The reduced update Kalman filter comprising (5.52–57) has been derived as an optimal approximation to the 2-D Kalaman scalar filter. The prediction part of the Kalman filter is left unchanged and the update is optimized under the constraint of updating only the nearest previously processed neighbors. Thus the update is optimal subject to the reduced update constraint but the prediction may not be constrained optimal. Such an algorithm will be called *weakly optimal.* When both the prediction and update are constrained optimal, the filter will be called *strongly optimal.*

Unfortunately the 2-D reduced update filter is not strongly optimal [5.4]. It can be shown that

$$E[s_1(m,n)|\hat{s}_1(m,n)]=\hat{s}_1(m,n)\,, \tag{5.58}$$

is both necessary and sufficient for the optimality of dynamical prediction. However (5.58) is not generally true for the reduced update filter. This is because as the local state vector $\hat{s}_1(m,n)$, the data on which the dynamical prediction is based, moves across the picture, old elements move from the inactive memory into the active or updated memory. For this reason, the elements of $\hat{s}_1(m,n)$ may not be jointly optimal in the sense of (5.58). Having seen this, one may construct a counter example to prove the nonoptimality of using the dynamical model for 2-D prediction. Thus the 2-D reduced update filter is not strongly optimal. Two questions then arise: 1) How does one modify the reduced update filter to be strongly optimal? 2) Is the reduced update filter approximately strongly optimal? These questions are answered next.

Modifications for Strong Optimality

The most straightforward modification for strong optimality is to calculate an optimal prediction vector

$$b(m,n)\triangleq E[\hat{s}_1(m,n)\hat{s}_1^{\mathrm{T}}(m,n)]^{-1}E[s(m+1,n)\hat{s}_1(m,n)]\,. \tag{5.59}$$

Then the resulting optimal prediction would be

$$\hat{s}(m+1,n)=b^{\mathrm{T}}(m,n)\hat{s}_1(m,n)=\sum_{\mathscr{R}_{\oplus+}} b_{k,l}^{(m,n)}\hat{s}^{(m,n)}(m-k,n-l)\,, \tag{5.60}$$

where the $b_{k,l}^{(m,n)}$ are the properly ordered elements of $b(m,n)$. The reduced update part of the algorithm would go through unchanged. However in order to calculate (5.59), one would have additionally to compute and update the concerned estimate covariances and cross-covariances between the estimate and the signal. While this would represent a significantly increased design burden, the computational complexity of the steady-state filter would clearly not be affected.

Approximate Strong Optimality

To demonstrate the approximate strong optimality of the unmodified reduced update filter, several examples were run using homogeneous data whose statistics exactly matched those of several (2×2)th order models. Steady-state reduced update filters were generated and used to filter the data. Running average estimates of the covariance matrix and cross-covariance vector of (5.59) were calculated over the 128×128 images (leaving out a ten pixel border on all sides). An optimal steady-state prediction vector $\boldsymbol{b}$ was thus determined. In a second filter run, this optimal $\boldsymbol{b}$ was used to predict $s(m+1,n)$ as a side calculation, and the mean square prediction error was tabulated. Also the mean square prediction error using the dynamical model $\boldsymbol{c}$ vector, was calculated. The dynamical and optimal prediction error variances are seen in Table 5.1. The prediction error values in the table show that the dynamical prediction is nearly optimal for these examples[3]. Figure 5.11 shows the percent excess prediction mean square error versus signal-to-noise ratio (SNR) for the three examples of Table 5.1.

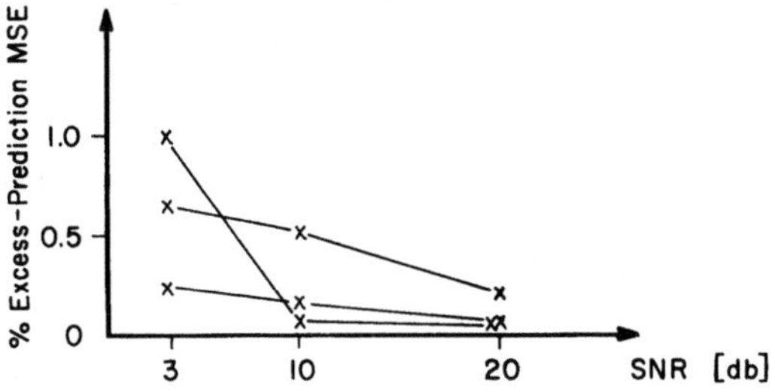

Fig. 5.11. Percent excess prediction mean square error using 2-D dynamical prediction

Table 5.1. Mean square prediction error summary (SNR = 3 db, $\sigma_v^2 = 0.5$)

Example	$\sigma^2(\boldsymbol{c})$	$\sigma^2(\boldsymbol{b})$
1	0.252	0.251
2	0.126	0.124
3	0.179	0.178

3 The reduced update region used for these examples is not exactly the local state defined earlier. It is the local support of the model mask at (m,n). Use of the exact local state leads to slightly higher errors for these examples. See [5.4] correction.

From these results follows the very important fact that for homogeneous data the Kalman reduced update filter can converge to a near optimal 2-D NSHP recursive filter for estimating the data. Now in one dimension, an exact filter could be obtained via Wiener's spectral design procedure. However, the same procedure in the 2-D case leads to an infinite-order filter, thus the optimal finite-order filter is not obtained. Hence the reduced update Kalman filter is also particularly attractive from the standpoint of design of spatially invariant filters for homogeneous random fields.

Order of Computation

Next we investigate the order of computation of this reduced update Kalman filter. We will consider each of (5.52–57) separately. First (5.52) will be $O(M^2)$ for an Mth order NSHP model. Equation (5.52) will thus be $O(M^2)$ for each (k,l). Since there are $O(MN)$ points in $\mathcal{S}_{\oplus+}^{(m,n)}$ we obtain $O(M^3N)$ for (5.53). Equation (5.54) is $O(M^2)$ as are (5.55, 56). Equation (5.57) is simple but has to be computed for each pair (i,j) and (k,l) with $(i,j)\in\mathcal{S}_{\oplus+}^{(m,n)}$, thus giving a computational total of $O(M^3N)$ as in (5.53). Summing up, we get the overall total computation per point as $O(M^3N)$. This is to be compared with $O(M^3N^3)$ for the general Kalman scalar filter and $O(M^2N^2)$ for the $(M\times M)$th order half-plane filter model.

The overall savings of a factor N^2 results from two simplifications. Firstly, the reduced update has reduced the orders of the matrices from $MN\times MN$ to $M^2\times MN$. Secondly, for the $(M\times M)$th order filter model, the scalar equations (5.52–57) only write the newly computed values at point (m,n). Equation (5.47) for example contains many more error covariance values than (5.53, 54). However, only the first row (and column) of $\boldsymbol{P}_b(m,n)$ actually changes. The other elements simply get shifted. The scalar equations were written with respect to a fixed reference, the origin in the data plane. Thus no shifting appears in that notation. Another way of looking at this is as follows. The $\boldsymbol{C}$ matrix is composed of 0's and 1's except for the first row. These 0's and 1's simply serve to shift the data in the state vector, one place down as each new element is put in. This computation can be avoided by using indirect addressing and simply writing the new element over the oldest element in storage. Then a pointer or indirect address can be used to keep track of the "head" of the vector. This is essentially the reason for not counting the shifting operations as they can be avoided with a minimal computational effort on many machines.

For the reasons outlined in the previous paragraph, (5.52–57) not only provide a convenient way to perceive the two-dimensional nature of the reduced update filter but also present its essential computational aspects. The main computational load was observed to be in (5.53, 57). Equation (5.53) computes the error covariance between the "predicted point" (m,n) and the previous estimates in the filters global state. Experience suggests this will be peaked at the point (m,n) with rapid decay with distance from (m,n). Thus it is a resonable approximation to only compute (5.53) in a fixed size region including

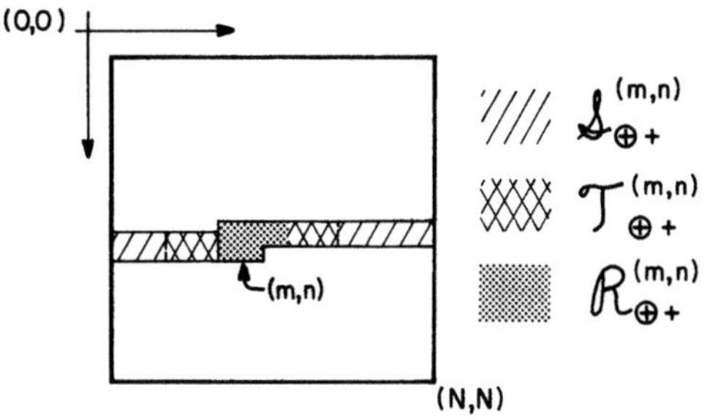

Fig. 5.12. Region assignment for approximate reduced update filter

$\mathscr{R}^{(m,n)}_{\oplus+}$. This reasoning can also be applied to (5.57), where (k,l) would be restricted to a region significantly smaller than $\mathscr{S}^{(m,n)}_{\oplus+}$ and of fixed size for increasing N. Calling such a region $\mathscr{T}^{(m,n)}_{\oplus+}$ we can rewrite (5.53, 57) as,

$$R_b^{(m,n)}(m,n;k,l)=\sum c_{op}R^{(m-1,n)}(m-o,n-p;k,l)\,, \quad (k,l)\in\mathscr{T}^{(m,n)}_{\oplus+}\,, \tag{5.61}$$

$$R_a^{(m,n)}(i,j;k,l)=R_b^{(m,n)}(i,j;k,l)-k^{(m,n)}(m-i,n-j)R_b^{(m,n)}(m,n;k,l)\,,$$
$$\text{for} \quad (i,j)\in\mathscr{R}^{(m,n)}_{\oplus+}\,;\ (k,l)\in\mathscr{T}^{(m,n)}_{\oplus+}\,. \tag{5.62}$$

These approximate reduced update equations reduce the order of the computations to $O(M^4)$, a constant with respect to N. Figure 5.12 sets forth the desired region assignment. The adverse effects of substituting $\mathscr{T}$ for $\mathscr{S}$ have been found minimal for most pictures. This is so both for the reasons mentioned previously and also because (5.55), the Kalman gain calculation, does not directly make use of points outside $\mathscr{R}^{(m,n)}_{\oplus+}$. Thus the effect of this truncation would be at worst second order.

If we look at memory requirements, we find that they are dominated by the need to store the $\boldsymbol{R}_a$ error covariances. For the 2-D Kalman scalar filter, this storage will be (M^2N^2). For the reduced update filter we get the same amount of storage, only a small portion of which is accessed at each point. For the approximate reduced update filter, we obtain $O(M^3N)$. A little thought reveals that this is the minimal possible dependence on N, i.e., linear, for a spatially varying processor[4]. Thus, we cannot hope for further improvements here. However, for homogeneous regions, these storage requirements can be greatly relaxed. Then one only has to run the processor over a much smaller region to obtain near convergence to the steady-state filter. Subsequently only (5.52, 56) need be computed at a substantially lower amount of computational effort.

Example

In an initial experiment with the reduced update filter, the noisy picture of Fig. 5.9b was modeled with a (3×3)th order $\oplus+$ NSHP model and processed

[4] To see this note that the processor must minimally have access to the error variances for the entire previous line.

Fig. 5.13. Reduced update filtering example, improvement is 8.3 db

by the approximate reduced Kalman filter for various sizes of $\mathscr{T}_{\oplus +}$. In particular region half-widths of 3, 5, 7, and 9 were tried. The results were approximately the same in all cases indicating both that the error decorrelates fast for this model and that the filter is fairly robust with respect to this type of error. The measured rms improvement was 8.3 db with output shown as Fig. 5.13. Convergence was obtained in 10 lines or less with a run time of 40 s on the CDC 7600. This run time could be greatly shortened if the filter were updated only for the first ten columns in the first ten lines instead of the entire first ten lines. The error covariance values of a sampling of the columns indicate that such a change in the algorithm would have negligible effect on the final result. This is because the chosen covariance model, as is typical of images, does not show substantial correlation over distances on the order of 10 or more pixels.

These results compare to those obtained by *Jain* [5.6]. However his 2-D recursive algorithms are costly computationally in that no reduced update constraints are provided. Comparatively, his transform based algorithms are more efficient, but require large sine transforms which could impose large I/0 costs on minicomputers.

5.3.3 Extension to Image Restoration [5.13]

In image restoration the observation equation (5.14) is modified to reflect a transformation on the signal, which specialized to the linear shift-invariant (LSI) case becomes

$$r(m,n)=\sum_{kl} h_{kl} s(m-k,n-l)+v(m,n) \ . \tag{5.63}$$

For a particular support of the distorting system h, (5.63) can be put into the form (5.17). Namely the support of h must be restricted to the global state region. If additionally h is only nonzero on the support of c, then h can be represented by $\boldsymbol{h}$ with $\boldsymbol{h}_2=\boldsymbol{0}$, and hence is in the class of problems to which the reduced update vector equations (5.47–51) apply. Such an h is an $(M\times M)$th order $\oplus+$ nonrecursive or FIR filter. This observation model can be obtained in various ways, e.g., least squares estimation [5.14], maximum-likelihood methods [5.15], and 2-D spectral modeling [5.16]. The FIR constraint on h is particularly appropriate for motion blurs and finite size apperture effects. Zero phase transformations may be incorporated in the model by tacitly agreeing to accept a small known shift in the output estimate.

Assuming h is an NSHP FIR filter of order $(M\times M)$ or less, we have (5.63) equivalent to (5.14) with $\boldsymbol{h}_2=0$ and

$$\boldsymbol{h}_1^{\mathrm{T}}=[h_{0,0},\dots,h_{-M,0};h_{M,-1},\dots,h_{-M,-1};\dots;h_{M,-M},\dots,h_{-M,-M}]\,.$$

Then the 2-D reduced update filter applies to this problem. Converting to the computationally more efficient scalar equations, we obtain the following theorem.

Theorem 3. The scalar reduced update filter equations for the signal model of (5.4) and the observations (5.63) where h is FIR with support contained on the support of c, is given as,

$$\hat{s}_b^{(m,n)}(m,n)=\sum_{kl}c_{kl}\hat{s}_a^{(m-1,n)}(m-k,n-l);\quad (m,n)\in[1,N]^2\,, \tag{5.64}$$

$$\hat{s}_a^{(m,n)}(i,j)=\hat{s}_b^{(m,n)}(i,j)+k^{(m,n)}(m-i,n-j)$$
$$\cdot\left[r(m,n)-\sum_{kl}h_{kl}\hat{s}_b^{(m,n)}(m-k,n-l)\right];\ (i,j)\in\mathscr{R}_{\oplus+}^{(m,n)}\,, \tag{5.65}$$

$$R_b^{(m,n)}(m,n;k,l)=\sum_{op}c_{op}R_a^{(m-1,n)}(m-o,n-p;k,l)\,,\qquad (k,l)\in\mathscr{S}_{\oplus+}^{(m,n)}\,, \tag{5.53}$$

$$R_b^{(m,n)}(m,n;m,n)=\sum_{kl}c_{kl}R_b^{(m,n)}(m,n;m-k,n-l)+\sigma_w^2\,, \tag{5.54}$$

$$k^{(m,n)}(i,j)=\frac{\sum_{kl}h_{kl}R_b^{(m,n)}(m-k,n-l;m-i,n-j)}{\sum_{kl}\sum_{op}h_{kl}h_{op}R_b^{(m,n)}(m-k,n-l;m-o,n-p)+\sigma_v^2};\quad (i,j)\in\mathscr{R}_{\oplus+}\,, \tag{5.66}$$

$$R_a^{(m,n)}(i,j;k,l)=R_b^{(m,n)}(i,j;k,l)-k^{(m,n)}(m-i,n-j)$$
$$\cdot\sum_{op}h_{m-o,n-p}R_b^{(m,n)}(m-o,n-p;k,l)\,;$$
$$(i,j)\in\mathscr{R}_{\oplus+}^{(m,n)},\quad (k,l)\in\mathscr{S}_{\oplus+}^{(m,n)}\,. \tag{5.67}$$

These Eq. (5.64–67) constitute the reduced update filter for deconvolution problems. The equations are seen to be very similar to (5.52–57) for estimation problems when signal distortion is not present. In fact the equations associated with prediction are unchanged and those associated with updating are modified only by convolution with the kernal h in appropriate places. This is analogous to the one-dimensional case.

Examples

For experimental verification blurred images were generated using uniform spread functions of finite extent. These point-spread functions could correspond to either finite size aperture effects (4×4 model) or to linear motion blur (7×1 model). Measurement noise $w(n)$ was then added to the blurred images using a Gaussian white noise generator. For simulation purposes, a signal-to-noise ratio of 40 db was setup with signal variance equal to unity and mean of zero. The noisy blurred images are shown in Fig. 5.14.

These images are processed using the deconvolution reduced update algorithm. For estimation, a stationary $\oplus +$ NSHP filter of appropriate order [i.e., a second-order filter for (3×3) model and a third-order filter for (4×4) and (7×1) models] was considered. The filter parameters were derived using a least-squares linear prediction approach [5.17]. In order to improve the program efficiency and to reduce the memory storage, the error covariance processing and gain calculation part of the reduced update filter were decoupled from the picture processing part. The error covariance processing program was first used to "design" the filter by iterating towards convergence. The steady state was declared "reached" after 10 rows and 20 columns for the second-order filter and after 15 rows and 25 columns for the third-order filter. In this design, the approximate reduced update of the error covariance matrix was limited to the $\mathscr{T}_{\oplus +}$ region of half-width equal to 5 for the second-order filter and to 7 for the third-order one. The noisy blurred images were then processed using the resulting steady-state filters.

a b

Fig. 5.14. FIR blurred images (noise level is -40 db): (a) 4×4 blur; (b) 7×1 blur

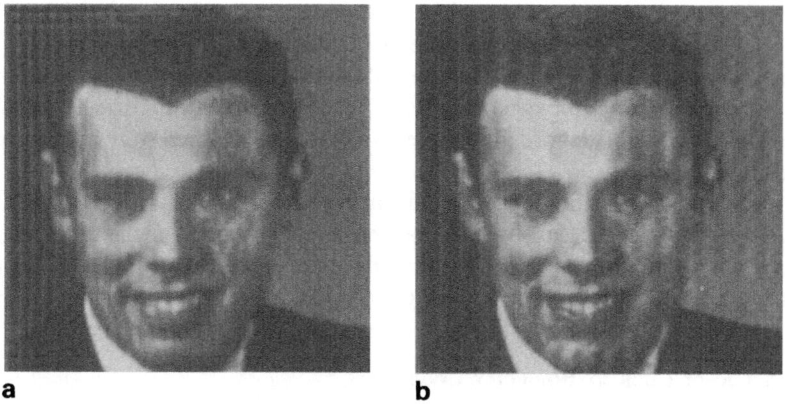

Fig. 5.15. Recursive restoration examples: (**a**) 4×4 blur; (**b**) 7×1 blur

Table 5.2. Mean square error and SNR improvement

Image description $\sigma_s^2 = 1.000$	MSE (input)	MSE (output)	Improvement SNR (db)
(3×3) model	0.0660	0.0074	9.5
(4×4) model	0.1138	0.0258	6.5
(7×1) model	0.0804	0.0319	4.1

The boundary condition set was taken as the noisy values in an M-pixel border on the top and sides of the noisy images. Thus the initial error covariance becomes that of the noise which is white. The fact that this does not correspond to the optimal *a priori* estimate causes a slight suboptimality near the boundary. The actual filtering is done with the steady-state reduced update filter, however, which is not affected by the assumed initial boundary values. The advantage of using these boundary conditions, besides their simplicity, lies in the fact that the corresponding initial error variances will be in a neighborhood of their final steady-state values. This will yield good numerical performance in the presence of the truncation errors which are generated by the approximate reduced update procedure.

The processed images are shown in Fig. 5.15. The results, in the form of measured SNR improvement are shown in Table 5.2.

As seen from this table, there is a considerable amount of improvement in the mean square error. This improvement is highest for the (3×3) point-spread function model (not shown). The subjective improvement is very good, making the restored image visually nearly the same as the original. In the case of the (7×1) and (4×4) point-spread functions, both the numerical and subjective results are less than perfect. Ringing arises due to the signal and noise exciting peaks in the response of the deconvolution filter. Subjectively the results are still quite dramatic in the increased sharpness of the restored images. Some further subjective improvement could probably be obtained by optimizing over the design signal-to-noise ratio parameter to maximize this improvement.

One caution in interpreting these results is that, since this is a simulation, the point-spread function is known exactly. If it were not known exactly, performance would be degraded. Further work will be devoted to restoring real blurred images.

As in Sect. 5.3.2, these deconvolution versions of the reduced update equations apply for the case of zero boundary conditions on the zero-mean field $s(m,n)$. In many applications, including image restoration, it is more natural to consider random boundary conditions, as done in the next section.

5.3.4 A Closer Look at Boundary Conditions

This section focuses on the 2-D boundary condition problem for spatial Kalman estimators [5.13]. It starts off with a review of the role of boundary conditions in 1-D recursive estimation.

The initial condition problem for 1-D Kalman filtering is to select an optimal initial estimate of the state, along with its associated initial error covariance. Often this estimate is simply the mean of the signal process as would be appropriate, for example, when no *a priori* information is available. When the data set is finite, a boundary value problem consisting of both initial and final conditions must in general be considered (see Fig. 5.16a). If one has a finite length section of an infinite length data set, the causal filtering problem ignores any final conditions, thus enabling one to reduce the general boundary condition problem to an initial value problem. Hence the two-point boundary nature of a finite length data set can be ignored in the 1-D case by the expedient of causal filtering. The general 2-D boundary condition set is shown in Fig. 5.16b. Unlike its 1-D counterpart, the boundary condition problem in 2-D does not degenerate to an initial condition problem for causal filtering. This is because the data scan causes the boundaries to be encountered repetatively in the course of the filtering, at the beginning and end of each scan line. Thus, in addition to initial conditions at the top and final conditions on the bottom, the majority of the boundary consists of points that cannot be so catagorized and

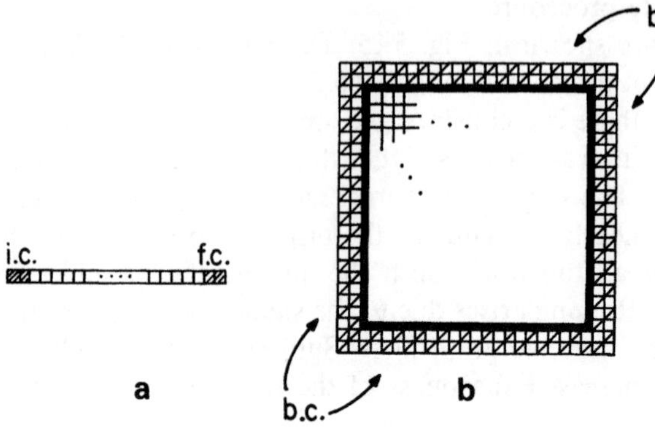

Fig. 5.16. Boundary value problems: (**a**) one-dimensional; (**b**) two-dimensional

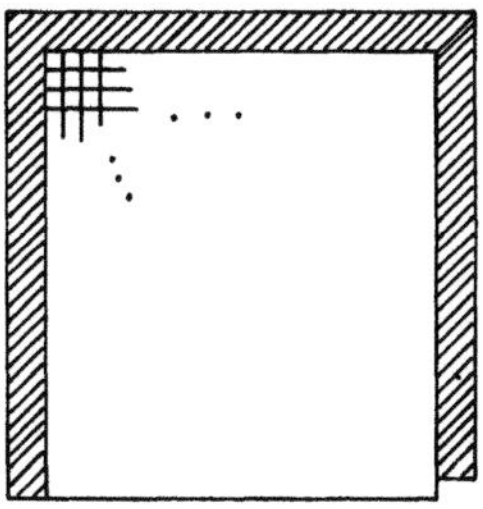

Fig. 5.17. Boundary condition set for 2-D causal recursive estimation

therefore has no 1-D counterpart. Therefore, unlike the 1-D case, a causal 2-D filter cannot ignore the boundary condition nature of the problem.

Figure 5.17 shows the minimum boundary condition set to uniquely specify the output of an NSHP recursive estimator or filter on a square region[5]. The boundary condition set is much more than an initial condition set.

In performing the recursive estimation, for random boundary conditions, one has the following options:

a) update the boundary conditions as encountered, or,

b) do not update boundary conditions.

In this regard, various questions arise. How does one do 2-D Kalman filtering in the presence of these random boundary conditions? How does the decision to update or not update the boundary conditions affect the optimality of the 2-D Kalman filter? Should the global and local states be modified for the case of random boundary conditions? We will attempt to answer these questions in the remainder of this section.

We assume in this section that the boundary conditions are sets of random variables, since if the boundary conditions are deterministic only straightforward modifications of methods already presented are needed. In this case, the method presented in [5.4] and Sects. 5.3.2, 3 where the boundary conditions are taken to be zero, can be modified by simply inserting the known boundary values into the equations when needed. This is completely analogous to the 1-D Kalman filter with deterministic input, and in fact through the raster scanning, is equivalent to it. In this case, the state vector consists of the last M lines up to and including the current point for an $(M \times M)$th order model.

The case of random boundary conditions is more realistic, especially for images of objects extending beyond the field of view. Random boundary conditions are also necessary to properly characterize a square region obtained from a homogeneous random field. Taking option (a) above, we must modify the 2-D Kalman filter to update the boundary conditions as encountered. To do this, it will be necessary to modify the global state to include those boundary

5 We note parenthetically that a quarter-plane recursive estimator or filter would require the boundary conditions of Fig. 5.17, with the exception of those on the right-hand side. This somewhat simpler situation is qualitatively unchanged from the general NSHP case to be presented here.

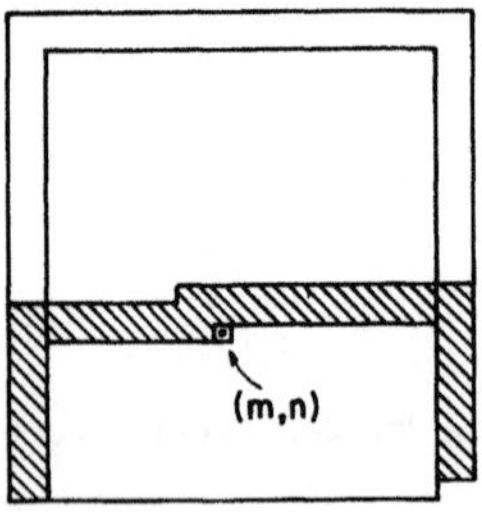

Fig. 5.18. The augmented global state region at pixel (m, n)

values which are to be updated at any given point. To completely characterize future response, the global state at (m, n) must have support as shown in Fig. 5.18. This new global state is the global state of [5.4] and Sects. 5.3.2, 3 augmented by the set of boundary values to be encountered in the future. This augmented global state will permit updating those boundary values which will be involved in future estimates. The 2-D Kalman filter for this augmented global state is derived, as before, through a raster scanning argument converting the 2-D problem into an equivalent 1-D problem. For all those points in $A^N = [1, N]^2$, we have

$$\boldsymbol{s}'(m,n) = \boldsymbol{C}'\boldsymbol{s}'(m-1,n) + \boldsymbol{w}'(m,n) \tag{5.68}$$

where the augmented global state vector $\boldsymbol{s}'$ is given as

$$\boldsymbol{s}'(m,n) \triangleq \begin{bmatrix} \boldsymbol{s}(m,n) \\ \hline \boldsymbol{b}(m,n) \end{bmatrix}, \tag{5.69}$$

where $\boldsymbol{b}(m, n)$ is a vector of boundary conditions as specified in Fig. 5.18. Actually no boundary values are computed by (5.68), the partition of the $\boldsymbol{C}'$ matrix being such that the $\boldsymbol{b}(m, n)$ undergo an identity transformation. Their presence only permits the calculation of $\boldsymbol{s}(m, n)$ near the beginning and end of lines, where the model support overlaps the boundary condition set. The augmented state thus does not change the prediction step of the 2-D Kalman filter except that the presence of the boundary values in the global state will allow the calculation of error cross-covariances between the data and boundary condition values. The Kalman filter, in turn, will then be able to calculate optimal updates of both the boundary values and the data values. This can be summarized as follows.

Theorem 4. The solution of the linear causal estimation problem using the 2-D Kalman filter, for raster scanned scalar observations on A^N and random boundary values, requires a global state augmented to include all boundary values to be needed in the future.

Proof. The 2-D dynamical prediction requires boundary value estimates when near a side of A^N. To be causally optimum, these values must be conditional means given the past data. Therefore those boundary value estimates to be used in the future, must be updated at each point or predicted when first needed. However, it is shown in [5.13] that the required boundary value estimate is not a linear combination of elements of the unaugmented global state. This implies the global state must be augmented to include these needed boundary values. Thus the augmented global state of (5.68) and Fig. 5.18 is of the minimal size at each point.

Note that the global state reduces in size as the filter progresses down the image. However this presents no particular problem with the mathematics, since the Kalman filter is recursively derived from point to point. If we define the σ field of past and present observations $\mathscr{F}_{m,n}^N \triangleq \{r|(k,l) \in (\text{past and present}) \cap A^N\}$, we find the following Corollary to Theorem 4.

Corollary. The 2-D Kalman filter estimate, based on using the augmented state vector, is the causal conditional mean estimate

$$\hat{s}_a(m,n) = \boldsymbol{E}[s(m,n)|\mathscr{F}_{m,n}^N] .$$

The best causal estimate based on $\mathscr{F}_{m,n}^N$ can be shown to be asymptotic to the best causal estimate based on $\mathscr{F}_{m,n}^\infty$. Thus if we let $\hat{s}(m,n) \triangleq E[s(m,n)|\mathscr{F}_{m,n}^\infty]$ we can show the following result:

Theorem 5. Let $\boldsymbol{E}|s(m,n)| < \infty$. Then the 2-D Kalman filter, based on the augmented state vector, is asymptotically optimum as

$$A^N \to A^\infty = \mathscr{R}_{++} \triangleq \{m \geqq 0, n \geqq 0\} ,$$

i.e.,

$$\hat{s}_a(m,n) \to \boldsymbol{E}[s(m,n)|\mathscr{F}_{m,n}^\infty] .$$

Proof. From the Corollary to Theorem 4,

$$\hat{s}_a(m,n) = \boldsymbol{E}[s(m,n)|\mathscr{F}_{m,n}^N] .$$

So since

$$\mathscr{F}_{m,n}^N \subset \mathscr{F}_{m,n}^{N+1} \subset \ldots \subset \mathscr{F}_{m,n}^\infty ,$$

we have by a standard convergence theorem [5.18],

$$\boldsymbol{E}[s(m,n)|\mathscr{F}_{m,n}^N] \to \boldsymbol{E}[s(m,n)|\mathscr{F}_{m,n}^\infty] .$$

Thus we generate finite dimensional approximations to the infinite dimensional recursive estimation on $\mathscr{R}_{++}$.

Reduced Update Filter

The 2-D Kalman filter, while optimal in the causal sense for the assumed total ordering, is computationally intractable for reasonably sized images as pointed out in Sect. 5.3.1. The 2-D reduced update filter limits updates to the nearby points, supposed to be of high correlation, and omits wasteful updating of distant points supposed to be only slightly correlated with the present observation. On deciding to update only a part of the augmented 2-D state at any point, the derivation goes through analogously to the deterministic boundary case. Thus we obtain equations of the same form as (5.47–51), where s_1 is the updated part of the augmented state vector and s_2 is the not-updated part. For example, if taking option (b) above, we decide to not update the boundary conditions, we include present and future boundary values in $s_2(m,n)$, so that their updates will be omitted. At the same time, we can take $s_1(m,n)$ to only include points nearby the present observation to get a computationally effective filter which is weakly optimal and has been found experimentally to be nearly strongly optimal in several examples [5.4].

The scalar equations for this reduced update filter on the augmented state are the same as (5.52–57) with the exception that the range of indices changes slightly to accommodate the enlarged state. By letting the 2-D lattice extend out into the boundaries we can represent both the boundary values and nonboundary values by the same symbols. Taking the filtering equations first, (5.52) would remain unchanged with the understanding that boundary estimates be inserted into the *rhs* when $\mathcal{R}_{\oplus+}^{(m,n)}$ overlaps the boundary. Equation (5.56) would consist of updates of both data estimates and boundary estimates. The gain equation (5.55) and error covariance equations (5.53, 54, 57) would be changed accordingly. Notationally, all that need be done is to define $\mathcal{U}_{\oplus+}^{(m,n)}$ as the update region on both the data and boundary and $\mathcal{S}_{\oplus+}^{(m,n)}$ as the augmented state region. Then (5.52–57) hold exactly as written with $\mathcal{U}_{\oplus+}^{(m,n)}$ substituted for $\mathcal{R}_{\oplus+}^{(m,n)}$. Such an update region is depicted in Fig. 5.19 for three points on a scan line, corresponding to the first point $m=0$, a generic point m_o, and the last point $m=N$. As before, the filter gain computation then becomes very efficient, however the error covariance computation per point is still of order $O(N)$. At this stage an approximate fixed size region $\mathcal{T}_{\oplus+}^{(m,n)}$ is substituted for $\mathcal{S}_{\oplus+}^{(m,n)}$ in the error covariance computations. The resulting approximate reduced update

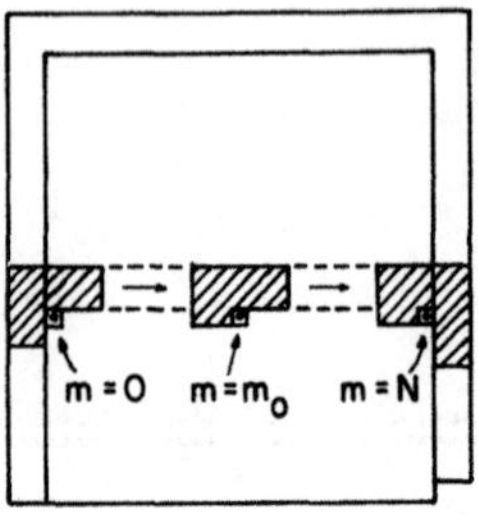

Fig. 5.19. Reduced update region for various values of m: $0, m_0, N$

algorithm will have fixed error covariance computation per point independent of image size. Moreover, the amount of error covariance storage can then be kept constant as the filter proceeds down the image. This is because, with updates restricted to the $\mathcal{T}$ region, as boundary points are encountered in $\mathcal{T}$ for the first time (at the beginning and end of lines), they have not been involved in previous updates. Therefore the error covariance values are initial conditions, which can be stored separately with a small amount of storage. Thus the use of the $\mathcal{T}$ region not only "bands" the error covariance matrix leading to reduced computation and storage requirements in a given row of the image, but it also computationally eliminates the algorithmic complexity of dealing with a global state of decreasing dimensions.

5.3.5 Multi-Model Estimation [5.19]

In the literature, various recursive techniques have been developed for one-dimensional estimation and detection of abrupt changes in the states [5.20, 21]. *Willsky* [5.20] considered a class of stochastic linear systems that are subject to jumps of unknown magnitude in the state occurring at unknown times. An adaptive filtering scheme is devised using Kalman-Bucy filtering and generalized likelihood ratio testing. *Lainiotis* [5.21] examined more general conditions for adaptive estimation which he called partition algorithms. These algorithms can be viewed as joint estimation and system identification algorithms where a linear discrete time stochastic dynamical system is parameterized by "L" time-invariant models. The solution will then consist of a bank of L Kalman filters each tuned to one of the models.

Though these techniques have been successfully applied in a wide variety of applications in 1-D situations, they are not suitable for image processing. Due to the scanning operation which converts the 2-D process into a 1-D process, these techniques fail to perform a true 2-D recursion and thus are insensitive to the variations in the nearest neighbors lying just above the current line. *Lebedev* and *Mirkin* [5.22] proposed a composite model of an image consisting of a finite number of classes of image fragments distinguished by the types of correlational links between the elements of the structure. This model was used for synthesizing a smoothing algorithm for minimum mean square error. The algorithm is nonrecursive in nature and not suitable for real-time application.

In this section we combine the approach given in [5.22] together with the reduced update Kalman filter for state estimation. The resulting spatially variant processing shows a considerable improvement over the LSI processing.

Multiple-Model Recursive Estimation

Let $\boldsymbol{s}_1(m,n)$ be the local state vector as shown in Fig. 5.10. Let $\boldsymbol{r}_1(m,n)$ and $\boldsymbol{v}_1(m,n)$ be the observation vector and white Gaussian noise vector, respectively, having the same support as that of the local state vector $\boldsymbol{s}_1(m,n)$. Note that the scalars $r(m, n)$ and $v(m, n)$ in (5.14) are the leading elements of the vectors $\boldsymbol{r}_1(m, n)$ and $\boldsymbol{v}_1(m, n)$, respectively. Let $\boldsymbol{E}\{\boldsymbol{v}\}=\boldsymbol{0}$ and $\boldsymbol{E}\{\boldsymbol{v}\boldsymbol{v}^{\mathrm{T}}\}=\boldsymbol{Q}_{\mathrm{v}}$.

Then from (5.14),

$$\boldsymbol{r}_1(m, n)=\boldsymbol{s}_1(m, n)+\boldsymbol{v}_1(m, n)\ . \tag{5.70}$$

Now we make the following hypotheses to arrive at the composite model description of an image.

1) There are L a priori known classes $\{\theta_j\}_{j=1}^{L}$ of image structures distinguished by the various correlational links between the elements of the local state vector.

2) The probability distribution of the classes $\{p(\theta_j)\}_{j=1}^{L}$ is known a priori.

3) The conditional distributional of the images for a given class is a Gaussian distribution, i.e.,

$$p(\boldsymbol{s}_1|\theta_j)\approx(2\pi)^{k/2}|\boldsymbol{R}_{\theta_j}|^{-1/2}\exp\left(-\boldsymbol{s}_1^{\mathrm{T}}\boldsymbol{R}_{\theta_j}^{-1}\boldsymbol{s}_1/2\right). \tag{5.71}$$

Hence in light of these hypotheses, the dynamical equation becomes

$$\boldsymbol{s}(m, n)=\boldsymbol{C}(m, n;\theta)\boldsymbol{s}(m-1, n)+\boldsymbol{w}(m, n) \tag{5.72}$$

where θ takes one of the L values at (m, n).

Now a bank of Kalman filters running in parallel and each designed based upon the statistics of one of the L models can be designed. The steady-state solutions, in scalar form, are given by

$$\hat{s}_b^{(m,n)}(m, n;\theta_j)=\sum_{kl}c_{k,l}(\theta_j)\hat{s}_a^{(m-1,n)}(m-k, n-l;\theta_j)\ , \tag{5.73}$$

$$\begin{aligned}\hat{s}^{(m,n)}(k, l;\theta_j)&=\hat{s}_b^{(m,n)}(k, l;\theta_j)+k(m-k, n-l;\theta_j)\cdot[r(m, n)-\hat{s}_b^{(m,n)}(m, n;\theta_j)]\ ,\\ &\quad (k, l)\in\mathscr{R}_{\oplus+}^{(m,n)}\quad \text{for}\quad j=1, 2, \ldots, L\ ,\end{aligned} \tag{5.74}$$

where the steady-state gain k is determined from the error covariance equations (5.54, 61, 62). The edge detection algorithm uses the noisy observations r to compute the a posteriori probability of the acting model being θ_j for each j. Using Bayes rule, we obtain,

$$P(\theta_j|\boldsymbol{r}_1)=\frac{p(\boldsymbol{r}_1|\theta_j)P(\theta_j)}{\sum\limits_{j=1}^{L}p(\boldsymbol{r}_1|\theta_j)P(\theta_j)}\qquad j=1, \ldots, L\ . \tag{5.75}$$

It now remains to compute the conditional distribution of the received vector $\boldsymbol{r}_1$ conditioned on model θ_j from,

$$p(\boldsymbol{r}_1|\theta_j)=p[\boldsymbol{s}_1(\theta_j)+\boldsymbol{v}_1]\ . \tag{5.76}$$

Using the Gaussian nature of $s_1(\theta_i)$ and v_1 and using the fact that they are independent we get: $r_1(m,n)$ conditioned on θ_j is Gaussian with mean zero and covariance $R_{\theta_j}+Q_v$, for each j.

Hence,

$$p(r_1|\theta_j)=(2\pi)^{-k/2}|R_{\theta_j}+Q_v|^{-1/2}\exp[-1/2r_1^{\mathrm{T}}(R_{\theta_j}+Q_v)^{-1}r_1]. \quad (5.77)$$

Using the above equations, we get,

$$P(\theta_j|r_1)=\frac{P(\theta_j)|R_{\theta_j}+Q_v|^{-1/2}\exp[-1/2r_1^{\mathrm{T}}(R_{\theta_j}+Q_v)^{-1}r_1]}{\sum_{j=1}^{L}P(\theta_j)|R_{\theta_j}+Q_v|^{-1/2}\exp[-1/2r_1^{\mathrm{T}}(R_{\theta_j}+Q_v)^{-1}r_1]}. \quad (5.78)$$

Then using the MAP decision rule, the switching logic simplifies to: select model θ_j iff

$$1/2r_1^{\mathrm{T}}(R_{\theta_j}+Q_v)^{-1}r_1+c_j\leqq 1/2r_1^{\mathrm{T}}(R_{\theta_i}+Q_v)^{-1}r_1+c_i, \quad i\neq j, \quad (5.79)$$

where

$$c_j\triangleq 1/2\log|R_{\theta_j}+Q_v|-\log P(\theta_j) \quad (5.80)$$

are constants which can be precomputed. Now (5.73, 74) along with (5.79) define a multiple-model recursive estimator (MMRE) which is shown in the block diagram of Fig. 5.20.

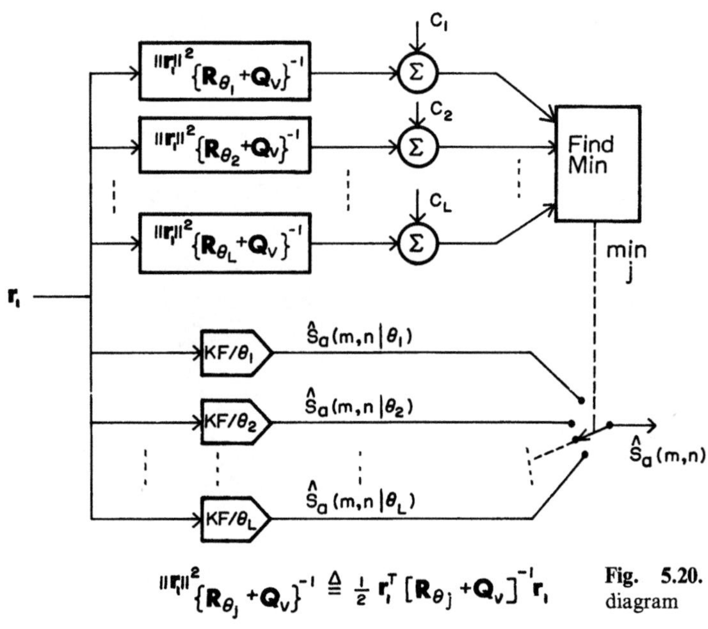

Fig. 5.20. MMRE block diagram

In the algorithm of (5.79), the decision rule needs the noisy local observations $\boldsymbol{r}_1$ which take into account the true 2-D nature of the problem thus helping to switch the model fast in a given situation. As such, this algorithm requires more processing compared to those in [5.20, 21]. Since each quadratic term is independent and uses the local observation vector $\boldsymbol{r}_1$, this algorithm has much parallelism. Also only one global state is stored and only one of the L Kalman filters is running at any given pixel.

Examples

The results of statistical measurements on sections of real images show that a good description of the images can be obtained with as few as five models. Four of the models correspond to the regions of the image with correlation directions of 0°, 45°, 90°, and 135° to the horizontal, for the images sampled and processed as rectangular arrays. The fifth model describes the regions of the image with

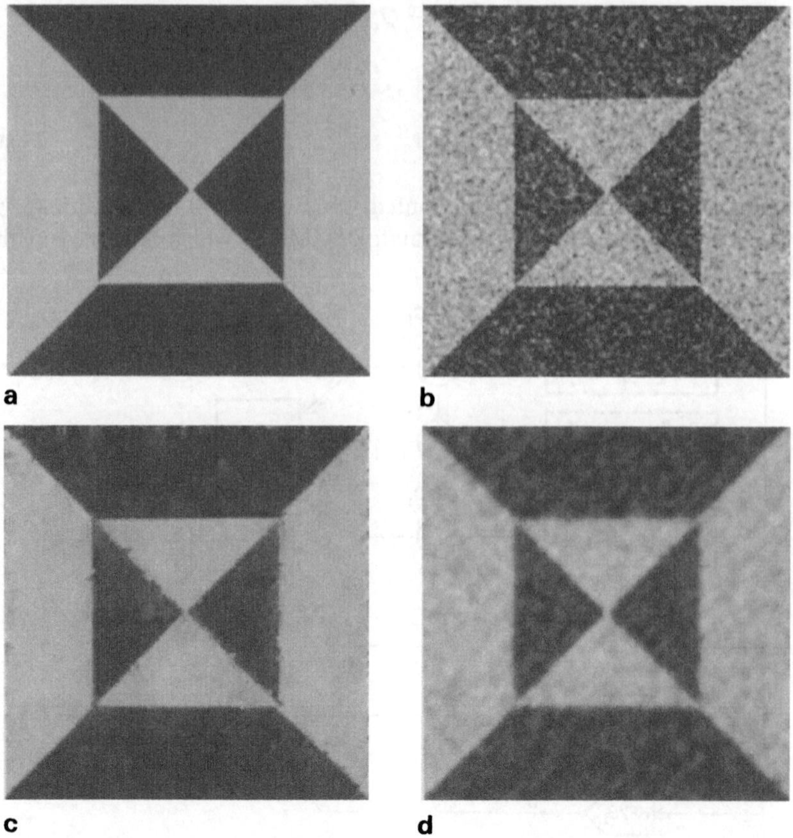

Fig. 5.21. MMRE example – geometrical image: **(a)** original; **(b)** noisy (3 db); **(c)** MMRE estimates; **(d)** LSI estimate

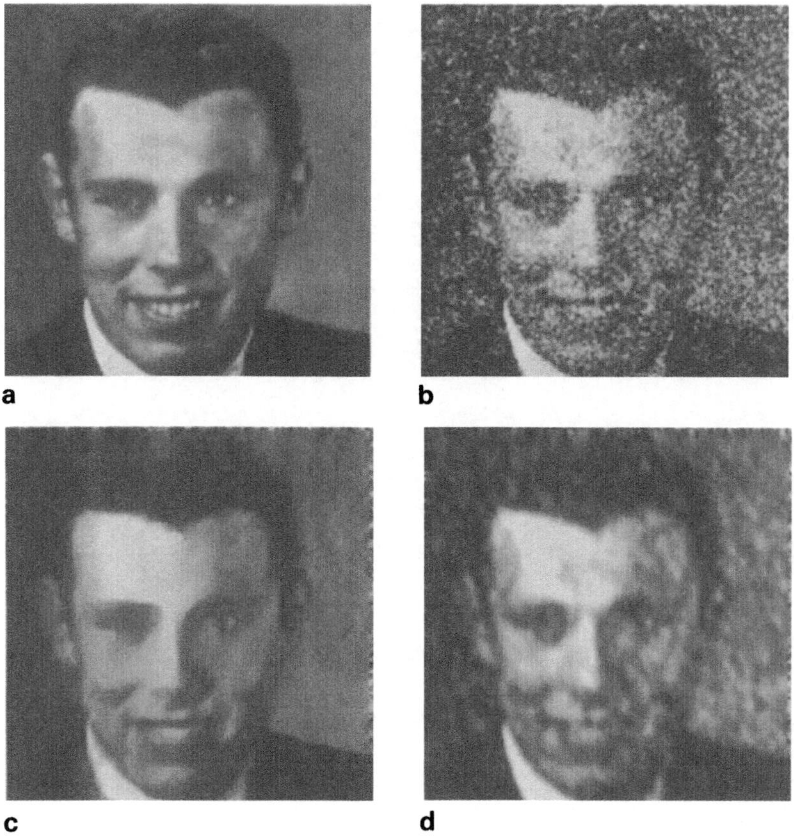

Fig. 5.22. MMRE example – face image: **(a)** original; **(b)** noisy (3 db); **(c)** MMRE estimates; **(d)** LSI estimate

isotropic structure. To test our algorithm, we used a synthetic image having the above-mentioned four predominant edges and an isotropic structure.

This synthetic image was a (128×128) image data field, quantized to 8 bits (256 gray-levels). The image is shown in Fig. 5.21a. For simulation purposes a signal-to-noise ratio of 3 db was set up with signal variance of unity and mean of zero. The noisy image is shown in Fig. 5.21b. For estimation, a second-order $\oplus +$ filter support was considered for all models. The appropriate steady-state filters were designed by using the reduced update error covariance equations. The noisy synthetic image was then processed using these five steady-state filters together with the switching algorithm of (5.79). The processed image is shown in Fig. 5.21c. For the purpose of comparison, the noisy image was also processed using the LSI or stationary filter whose parameters were identified using the space-variant linear prediction method, this is shown in Fig. 5.21d.

Comparison of Fig. 5.21c and d shows that multi-model recursive estimation (MMRE) gives a result with sharp edges and less blurriness than does

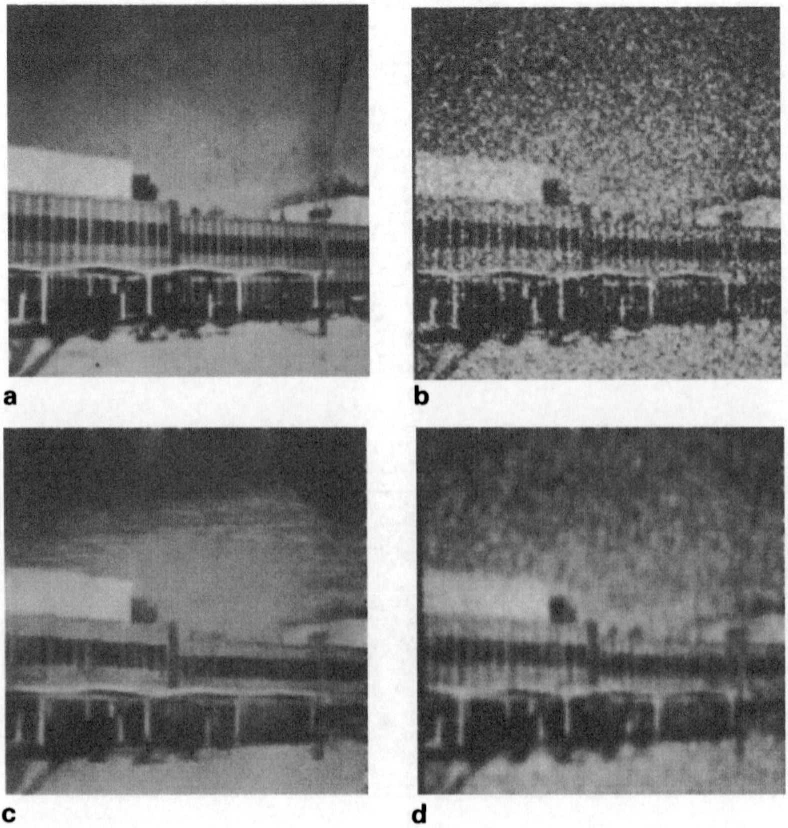

Fig. 5.23. MMRE example – building image; **(a)** original; **(b)** noisy (3 db); **(c)** MMRE estimate; **(d)** LSI estimate

the LSI estimate. The measured mean square error for Fig. 5.21c was 0.045 compared with 0.073 for the LSI. This considerable improvement in the numerical and the subjective quality shows that the composite model description gives a better fit to the image structure.

The same set of models was then used for the processing of noisy real images. The results are given in Fig. 5.22 for a face image, and in Fig. 5.23 for a building image. The mean square errors for the face were 0.097 for the MMRE and 0.063 for the LSI estimate. Thus the subjective and mean square error quality are reversed for these images. This almost surely results from using the geometrical image models for this face image.

Doubly Stochastic Gaussian Estimation

The above adaptive approach to the use of multiple models may be improved upon by introducing a spatial doubly stochastic Gaussian model as discussed in

Sect. 5.2.2. This will lead to a processor similar to the MMRE above but with the added advantage that the local state of the underlying Markov chain can alter the likelihood of the various model transitions and hopefully improve upon model switching decisions. This is indeed the case as will be shown below.

We consider a decision-directed approach. Accordingly assume that s and l are known for the past at pixel (m,n). The probability of $l(m,n)$ conditioned on the noisy observation of the local state $\boldsymbol{r}_1(m,n)$ can be written as

$$P[l(m,n)|\boldsymbol{r}_1(m,n), \text{past } l]$$

$$= \frac{p[\boldsymbol{r}_1(m,n)|\boldsymbol{l}(m,n), \text{past } l]\, P[l(m,n)|\boldsymbol{l}_1(m-1,n)]}{\sum\limits_{l=1}^{L} p[\boldsymbol{r}_1(m,n)|l, \text{past } l]\, P[l|\boldsymbol{l}_1(m-1,n)]}, \tag{5.81}$$

where use has been made of the conditional probability of the 2-D Markov chain. The function $P(\cdot|\cdot)$ then is just given by the corresponding row of the probability transition matrix of the chain. The conditional probability density of $\boldsymbol{r}_1(m,n)$ is Gaussian with mean zero and covariance determined by the past and present chain data l. If we assume all nearby points are in the same state l, then we may approximate the conditional density of $\boldsymbol{r}_1$ as $N(\boldsymbol{0}, \boldsymbol{R}_l+\boldsymbol{Q}_v)$. If we replace $\boldsymbol{l}_1(m-1,n)$ by its estimate $\hat{\boldsymbol{l}}_1(m-1,n)$, we obtain the following decision rule: Choose l such that

$$1/2\boldsymbol{r}_1^{\mathrm{T}}(\boldsymbol{R}_l+\boldsymbol{Q}_v)^{-1}\boldsymbol{r}_1 + c_l[\hat{\boldsymbol{l}}_1(m-1,n)]$$

$$= \underset{1\leqq k\leqq L}{\mathrm{Min}} \{1/2\boldsymbol{r}_1^{\mathrm{T}}(\boldsymbol{R}_k+\boldsymbol{Q}_v)^{-1}\boldsymbol{r}_1 + c_k[\hat{\boldsymbol{l}}_1(m-1,n)]\}, \tag{5.82}$$

where

$$c_l(\hat{\boldsymbol{l}}_1) \triangleq 1/2\log|\boldsymbol{R}_l+\boldsymbol{Q}_v| - \log P[l|\hat{\boldsymbol{l}}_1]. \tag{5.83}$$

This decision rule can be compared with (5.79, 80), the decision rule for the MMRE. The difference is the replacement of the a priori probabilities $P(\theta_i)$ by the conditional probabilities of the Markov chain given the decision-directed present state $\hat{\boldsymbol{l}}_1(m-1,n)$. If the recent decisions have been good, this has the effect of improving the probability of the correct decision on the l value at the present pixel.

The approximation to the conditional density of $\boldsymbol{r}_1(m,n)$ is convenient and can give good results. It can be heuristically justified by the following reasoning. Changes of chain state l are infrequent for the images under consideration. Thus for most pixels, the approximation that $\boldsymbol{r}_1$ is completely in one chain state should be quite good. In any event, the approximation allows the incorporation of some of the Markov chain information while still retaining the decision framework of the MMRE.

Examples

Figure 5.24 shows the estimate obtained by the doubly stochastic Gaussian algorithm consisting of the incorporation of the Markov chain into a decision-directed algorithm consisting of (5.82, 83) and the correspondingly modified versions of (5.73, 74). The noisy image is shown in Fig. 5.21b. The resulting mean square error is 0.041 which is approximately a 9% improvement over the MMRE estimate of Fig. 5.21. Subjectively the image is much improved.

Figure 5.25 shows similar results for the face image of Fig. 5.22b as input. The mean square error is 0.058 which is approximately an 8% improvement over the LSI estimate of Fig. 5.22d. The models here were identified from the noise free image using (5.94, 95) in Sect. 5.4.3 with a set of chain data provided by the decision logic in the MMRE algorithm. The estimate of Fig. 5.25 shows clearly increased sharpness and less noise in the flat regions than the LSI estimate.

Fig. 5.24. Doubly stochastic Gaussian example – geometric image

Fig. 5.25. Doubly stochastic Gaussian example – face image

a

b

Fig. 5.26a and b. 1×1 model: **(a)** LSI estimate; **(b)** doubly stochastic estimate

In Fig. 5.26 are shown processing results obtained using a (1×1)th order NSHP model and thus (1×1)th order reduced update filtering. Here the local state vector only contains 4 pixels in place of the 12 pixels in (2×2)th order model. Figure 5.26a shows the LSI estimate with a mean square error of 0.0895. Figure 5.26b shows the doubly stochastic estimate with a mean square error slightly less 0.0888. The subjective improvement of the doubly stochastic estimate is evident in the increased sharpness of the image. However there are more "false alarms" than in Fig. 5.25 due to the reduced support of the local state.

5.4 System Identification

This section focuses on the problem of determining the models for use in the reduced update Kalman filtering presented in Sect. 5.3. Three topics are treated: spectral modeling using 2-D spectral factorization to develop frequency domain NSHP models, linear predictive techniques for developing spatial domain models, and finally multiple-model techniques suitable for the maximum likelihood estimation of the models in either the adaptive multiple-model or the doubly stochastic Gaussian cases.

5.4.1 Spectral Modeling

The basic model of (5.4) can be viewed as a filter driven by the white noise w. The magnitude square transfer function of this filter is

$$|C(\mathrm{e}^{\mathrm{j}u}, \mathrm{e}^{\mathrm{j}v})|^{-2} .$$

If we take $S_s(\mathrm{e}^{\mathrm{j}u}, \mathrm{e}^{\mathrm{j}v})$ as the desired power density spectrum, then spectral modeling with the L^2 norm,

$$\|F\|_2 \triangleq \int_{-\pi}^{+\pi} \int_{-\pi}^{+\pi} |F(u,v)|^2 du dv ,$$

consists of finding the model $\{c_{kl}\}$ such that

$$\|S_s - \sigma_w^2 |C|^{-2}\|_2 \tag{5.84}$$

is minimized subject to the constraint that the factors be recursively stable. In the area of 2-D recursive filter design, various procedures have been developed for this nonlinear optimization problem with constraints [5.10, 23]. While globally optimum procedures do not generally exist, the optimization procedure described below, based on 2-D spectral factorization, is locally optimum.

Spectral Factorization

A procedure is developed in [5.9] for accomplishing two-dimensional spectral factorization for spectra satisfying mild conditions relating to continuity and boundness away from $+\infty$ and 0. This spectral factorization has several cannonic forms. The one of interest here results in NSHP $\oplus+$ factors which are recursively stable. The factorization is accomplished using a homomorphic transform involving the logarithm in the spatial frequency domain. The homomorphic transform is shown in Fig. 5.27, where the array $x(m,n)$ is homomorphically transformed into the cepstral array $\hat{x}(m,n)$ using the operator

$$\mathscr{H}=\mathscr{F}^{-1}\,\mathrm{Ln}\,\mathscr{F}\,, \tag{5.85}$$

where $\mathscr{F}$ is the 2-D Fourier transform and Ln is the pointwise natural logarithm operator. The array x is assumed to be a zero-phase element of l^1 with Fourier transform X satisfying $0<X<\infty$. The spectral factorization is then accomplished using $\mathscr{H}$ and $\mathscr{H}^{-1}$ as shown in Fig. 5.28. First the cepstrum $\hat{x}$ is computed from x. Then it is projected onto $\mathscr{R}_{\oplus+}$ yielding $\hat{x}_{\oplus+}$. Finally the $\hat{x}_{\oplus+}$ inverse transform $\mathscr{H}^{-1}$ yields the NSHP factor $x_{\oplus+}$. Similarly we may project onto $\mathscr{R}_{\ominus-}$ and obtain the factor $x_{\ominus-}\triangleq\mathscr{H}^{-1}(\hat{x}_{\ominus-})$. It is shown in [5.9] that these factors have the following properties for $x\in l^1$ and $X(u,v)>0$:

i) $x_{\oplus+}$ *(and* $x_{\ominus-}$) are recursively stable arrays,
ii) $x_{\oplus+}$ has NSHP support,
iii) $x_{\oplus+}(m,n)=x_{\ominus-}(-m,-n)$,
iv) $x=x_{\oplus+}*x_{\ominus-}$,
v) $X=|X_{\oplus+}|^2=X_{\oplus+}X_{\ominus-}$.

In general the exact factors will be infinite order, however, finite-order approximations can result in a specified small error. We can obtain such an approximation in the simplest case, by simply truncating the factors to finite support.

In [5.9] a window method is advanced as a simple way to design these factors. More elaborate and efficient design methods have been developed [5.10]. They essentially imbed the above factorization in an iterative nonlinear optimization loop with a weighted penalty function consisting of both frequency and spatial domain errors. The spatial domain error arises from the generally nonfinite support factors. Figure 5.29 shows an optimization algorithm approach to model design. An initial guess ($x^{(0)}_{\oplus+}$) is provided for the

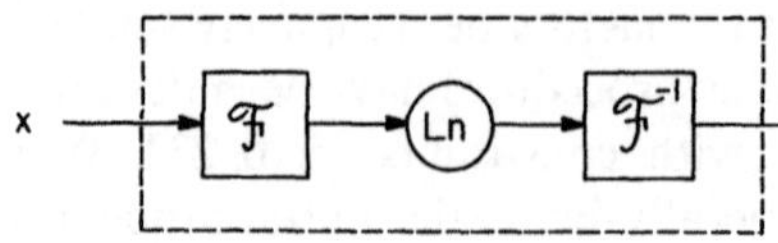

Fig. 5.27. A homomorphic transform

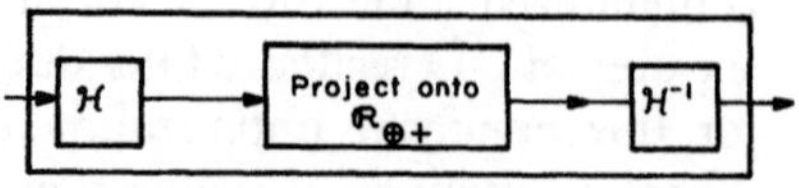

Fig. 5.28. Spectral factorization, NSHP $\oplus+$ canonical decomposition

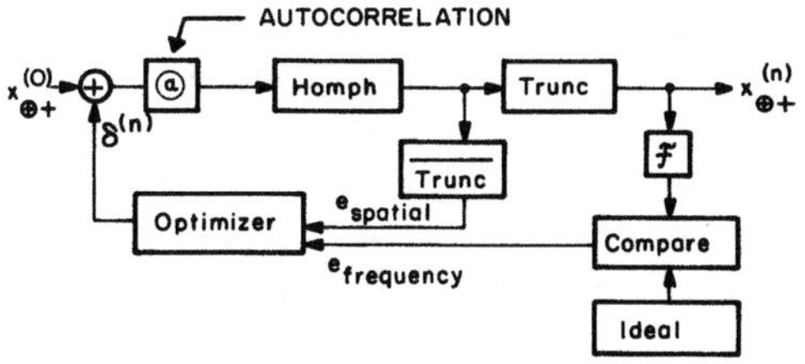

Fig. 5.29. Spectral design algorithm for $\oplus +$ AR modeling

factor $x_{\oplus +}$. The box "Homph" does the spectral factorization. The boxes Trunc and $\overline{\text{Trunc}}$ compute the finite order factor $x_{\oplus +}^{(n)}$ and the spatial domain error e_{spatial}. The spectrum of $x_{\oplus +}^{(n)}$ is compared with the design ideal spectrum to generate a frequency domain error $e_{\text{frequency}}$. The spatial and frequency domain errors then drive an optimizer to provide an increment $\delta^{(n)}$ for the iteration on the initial factor. More details and extensions to the algorithm as well as examples may be found in [5.10].

This identification technique was used with DFT based spectral estimation from the noise-free image to obtain the models for the girl's face in Figs. 5.9 and 13. The linear predictive least-sqares models generally have resulted in somewhat less filter estimation error. Subjectively the results are similar. Thus the linear prediction models, based on the space domain approximation, were used for the rest of the examples. Work on the spectral modelling is continuing with larger DFT sizes (i.e., greater than 32×32). A clear advantage of the spectral technique is that stability of the resulting recursive model is guaranteed.

5.4.2 Linear Predictive Techniques

In the linear prediction approach to identifying the NSHP model of (5.4) one minimizes the square residual

$$J_{m,n} = \sum_{(i,j)\in\Omega_{m,n}} [s(i,j) - \hat{c}^{\mathrm{T}} s_1(i,j)]^2 \, . \tag{5.86}$$

Here $s_1(i,j)$ is the local state vector at (i,j) and $\Omega_{m,n}$ is a local data history at pixel (m,n). The variance of the plant noise $w(m,n)$ is estimated as

$$\hat{\sigma}_w^2 = \frac{J_{m,n}}{N_{m,n}} \, , \tag{5.87}$$

where $N_{m,n}$ is the number of points in the data history $\Omega_{m,n}$. It is easily seen that if one assumes that the data are Gaussian and satisfy (5.4) for an assumed model order (M_1, M_2), then the above constitutes a maximum-likelihood estimate for the parameters c_{kl} and σ_w^2. The estimate for c becomes

$$\hat{c} = \left[\sum_{m,n} s_1 s_1^{\mathrm{T}}\right]^{-1} \left[\sum_{m,n} s_1 s\right] . \tag{5.88}$$

These equations assume that a noise-free sample of the random signal field is available for parameter estimation. This assumption is appropriate for the case of constant models where on-line adaptive parameter estimation is not required. However in the case of non-constant models where the spatial variation of the model parameters is to be determined, it is necessary to base the system identification on the noisy data, then (5.86, 87) become

$$J_{m,n} = \sum_{(i,j)\in\Omega_{m,n}} [r(i,j) - \hat{c}^{\mathrm{T}} \boldsymbol{r}_1(i,j)]^2 . \tag{5.89}$$

Here an estimate for $\boldsymbol{c}$ can be constructed as

$$\hat{\boldsymbol{c}} = \left[\sum_{m,n} (\boldsymbol{r}_1 \boldsymbol{r}_1^{\mathrm{T}} - \varrho\sigma_w^2 \boldsymbol{I})\right]^{-1} \left[\sum_{m,n} \boldsymbol{r}_1 r\right], \tag{5.90}$$

where $\varrho, 0 \leqq \varrho \leqq 1$, is a data dependent parameter chosen to partly remove the bias due to observation noise, while keeping the correlation matrix positive definite. A similar estimate for the variance is

$$\hat{\sigma}_w^2 = J_{m,n}/N_{m,n} - \beta(1 + \sum \hat{c}_{ij}^2)\sigma_v^2 , \tag{5.91}$$

where β, $0 \leqq \beta \leqq 1$, is chosen analogously to ϱ.

If the data s is homogeneous then the estimates given by (5.87, 88) are asymptotically consistent as the region $\Omega_{m,n}$ grows without bound. To investigate the rate of convergence, a homogeneous noise field was simulated using a (2×2)th order NSHP model identified from the face image. The regions $\Omega_{m,n}$ were chosen to be $K \times K$ blocks with the identification conducted at $(m, n) = (K, K)$. The results are shown in Table 5.3.

The results in the table point to a block size choice of 32×32 or larger for the ideal model used.

For the development of nonstationary models, more appropriate for real images, the data history region $\Omega_{m,n}$ must be kept small and local to track the parameter variations. To avoid statistical problems and still have small $\Omega_{m,n}$ regions, it was decided to try using (1×1)th order models. These models only have five parameters to identify versus the thirteen parameters of the (2×2)th model. In a subsequent experiment the 64×64 center segment of the face image was processed at SNR = 3 db using the reduced update Kalman filter and (1×1)th order models obtained by the least square identification procedure for $\Omega_{m,n}$ regions consisting of blocks of size 4×4, up to 64×64. The ϱ value used in (5.90) was $\varrho = 0$. Also the calculation of $\hat{\sigma}_w^2$ was performed on the noise-free data. The estimation error results are shown in Fig. 5.30.

The value $\varrho = 0$, chosen to avoid numerical problems in this adaptive procedure, causes a bias which shifts all the numerical results upward somewhat. Presently, more powerful methods are being investigated to remove this bias. Two resulting estimated images are seen in Fig. 5.31. The left image corresponds to $K = 8$ and the right image corresponds to $K = 64$, i.e., a constant

Table 5.3. Parameter estimates vs block size

K	$C_{2,2}$	$C_{1,2}$	$C_{0,2}$	$C_{-1,2}$	$C_{-2,2}$	$C_{2,1}$	$C_{1,1}$	$C_{0,1}$	$C_{-1,1}$	$C_{-2,1}$	$C_{2,0}$	$C_{1,0}$	$\hat{\sigma}_w^2$	ε^{2a}
True Values	0.138	−0.048	−0.096	0.03	−0.04	0.033	−0.61	0.839	−0.048	0.046	−0.231	0.968	0.0750	
4	−0.04	0.57	0.35	−0.27	0.29	0.20	−0.62	1.46	−1.49	1.16	−0.40	1.102	0.0114	1.43
8	0.15	−0.29	0.42	−0.51	0.27	0.17	−0.64	0.50	0.48	−0.28	−0.35	1.11	0.0554	1.57
16	0.13	−0.08	−0.00	−0.07	−0.008	0.07	−0.70	0.90	0.008	−0.02	−0.27	1.024	0.0773	0.89
32	0.16	−0.09	−0.05	−0.01	−0.02	−0.06	−0.54	0.84	−0.01	0.018	−0.15	0.92	0.0718	0.449
64	0.14	−0.07	−0.08	0.029	−0.05	−0.002	−0.56	0.85	−0.03	0.04	−0.19	0.927	0.0756	0.23
128	0.1375	−0.056	−0.089	0.023	−0.04	0.018	−0.59	0.841	−0.048	0.048	−0.214	0.956	−0.0749	0.118

[a] $\varepsilon^2 \triangleq \sum(\hat{c}_{ij} - c_{ij})^2$

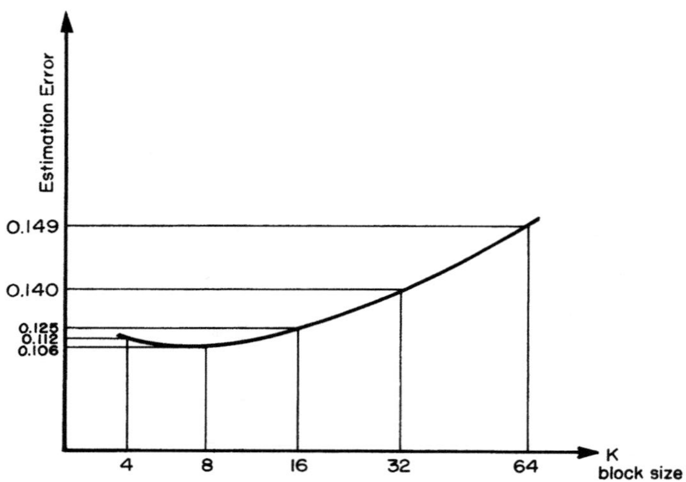

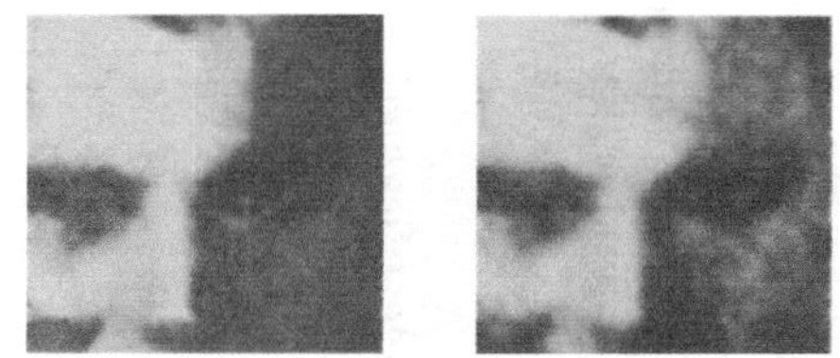

Fig. 5.31. Estimation results: left $-K=8$; right $-K=64$

Fig. 5.30. Reduced update filter estimation error versus identification blocksize (1 × 1th order ⊕+ model, SNR = 3 db) parameter model for this 64×64 block. One can detect an increased sharpness and reduced noise in the $K=8$ image along with some artifacts. More details may be found in [5.24].

In 2-D least squares parameter estimation using the linear prediction method, it may happen that the resulting autoregressive model is unstable [5.25] even though the relevant correlation matrices are positive definite. This is quite unlike the well-known case in one dimension. In the adaptive estimation work, it is not uncommon to obtain such unstable 2-D models, especially for low-order models when working with noise-free data. The identification can be "stabilized" by the expedient of adding a small amount of noise to the noise-free data. When working with noisy data, instability has not been encountered with $\varrho=0$ in (5.90). However, when one tries to set ϱ near to one to reduce the bias in the parameter estimate, the recursive stability requirement is an additional constraint that must be placed on the estimate $\hat{c}$.

5.4.3 Multiple-Model Identification

This section treats the system identification problem for the adaptive multiple models and the doubly stochastic Gaussian model introduced in Sect. 5.2.2 and utilized for image estimation in Sect. 5.3.5. The basic approach will involve maximizing the likelihood ratio of the joint field consisting of observations of both the data field $s(m,n)$ and the underlying Markov chain field $l(m,n)$. In this section it is assumed that a sample function noise-free signal field $s(m,n)$ is available for parameter identification. In practice this might take the form of a similar image or group of images. Note that such an approach is not reasonable for the spatially variant Gaussian case, i.e., without multiple models.

In the adaptive multiple-model case, it maybe reasongable to regard $\{l(m,n)\}$ as unknown parameters to be optimized over. The appropriate likelihood ratio becomes

$$LR(\sigma_1, \ldots, \sigma_2; c_1, \ldots, c_L) = \log p[\{s(m,n)\} | \{l(m,n)\}] = -\sum_{m,n} log\, \sigma_{l(m,n)}$$

$$-\sum_{m,n} \frac{1}{2\sigma^2_{l(m,n)}} [s(m,n) - c^{\mathrm{T}}_{l(m,n)} s_1(m-1,n)]^2 \,. \tag{5.92}$$

If $\{l(m,n)\}$ is known, then the maximum likelihood estimate can be given by a variation on linear least-squares prediction. If we define $L_l \triangleq \{(m,n) | l(m,n) = l\}$ and set $N_l \triangleq$ number of points in L_l, then

$$LR(\sigma_1, \ldots, \sigma_L; c_1, \ldots, c_L) = -\sum_{l=1}^{L} \mathrm{N}_l \log \sigma_l$$

$$-\sum_{L_l} \frac{1}{2\sigma_l^2} [s(m,n) - c_l^{\mathrm{T}} s_1(m-1,n)]^2 \,. \tag{5.93}$$

The maximum likelihood occurs at $\partial LR/\partial\sigma_l=0$ and $\nabla_{c_l}LR=\mathbf{0}$ for $l=1,\ldots,L$ which yields

$$\hat{c}_l=\left[\sum_{L_l}s_1(m,n)s_1^{\mathrm{T}}(m,n)\right]^{-1}\left[\sum_{L_l}s(m,n)s_1(m-1,n)\right], \tag{5.94}$$

$$\hat{\sigma}_l^2=N_l^{-1}\sum_{L_l}[s(m,n)-\hat{c}_l^{\mathrm{T}}s_1(m-1,n)]^2\,. \tag{5.95}$$

If one assumes $\{l(m,n)\}$ are not known, then one may treat them as unknown parameters and maximize over them. Then the following iterative procedure can be seen to increase the likelihood at each iteration.

Step 1: Guess $\{l(m,n)\}$.

Step 2: Solve (5.94, 95) to get $\hat{\sigma}_l$ and $\hat{c}_l$, $l=1,\ldots,L$.

Step 3: Change $\{l(m,n)\}$ one-by-one to maximize LR, using the parameter values of Step 2.

Step 4: Go to Step 2.

The utility of these algorithms, and variations on it, are presently under investigation. To cut down on computation, only portions of the image may be processed in iterating through steps 2–4.

Doubly Stochastic Gaussian Case

In the doubly stochastic case, the joint probability of $\{s(m,n)\}$ and $\{l(m,n)\}$ can be decomposed using the lemma of Sect. 5.2.2. Using the fact that $s(m,n)$, given $s_1(m-1,n)$ and $l(m,n)$, is Gaussian, we have

$$p[s(m,n)|l(m,n),s_1(m-1,n)]$$
$$=\frac{1}{\sqrt{2\pi\sigma_l^2}}\exp\left(-\left\{\frac{1}{2\sigma_{l(m,n)}^2}\;[s(m,n)-c^{\mathrm{T}}_{(m,n)}s_1(m-1,n)]^2\right\}\right). \tag{5.96}$$

Using (5.96) and the results of the lemma, we get

$$\log p[\{s(m,n)\},\{l(m,n)\}]$$
$$=\sum_{m,n}\log P[l(m,n)|\boldsymbol{l}_1(m-1,n)]-\sum_{m,n}\log\sigma_{l(m,n)}$$
$$-\sum_{m,n}\frac{1}{2\sigma_{l(m,n)}^2}[s(m,n)-c_{l(m,n)}^{\mathrm{T}}s_1(m-1,n)]^2\,. \tag{5.97}$$

Perhaps the simpliest case consists of observing both s and l with known L, $c_1,\ldots,c_L$ and $\sigma_1,\ldots,\sigma_L$. In this case we want to estimate the L^5 unknown parameters $\boldsymbol{P}[l|\boldsymbol{l}_1]$. The ML estimate is the well-known histogram estimate of $\boldsymbol{P}$.

If we know $\{s(m,n)\}$, $\{l(m,n)\}$ and L then the parameters $\boldsymbol{P}$ and $\boldsymbol{c}_1, \ldots, \boldsymbol{c}_L$ and $\sigma_1, \ldots, \sigma_2$ can be determined easily. The likelihood separates and $\hat{\boldsymbol{P}}$ is given by the histogram estimate and the $\hat{\boldsymbol{c}}_l$, σ_l estimates are provided by least squares linear prediction as in (5.94, 95).

If only $\{s(m,n)\}$ and L are known then the following iterative scheme is proposed:

Step 1: Guess $\{l(m,n)\}$.

Step 2: Find $\hat{\boldsymbol{P}}$ and $\hat{\sigma}_l$, $\hat{\boldsymbol{c}}_l$ as above, $l=1, \ldots, L$.

Step 3: Change $\{l(m,n)\}$ sequentially to maximize LR using the parameter values of Step 2.

Step 4: Go to Step 2.

Step 3 could be performed using the methods of maximum likelihood sequence estimation. Most notably the (M, L) algorithm of *Anderson* and *Bodie* [5.26] could likely be adapted for this purpose. Work in this area is ongoing.

5.5 Conclusions

This chapter has discussed the 2-D Kalman filtering methods of recursive estimation based on both vector or strip and scalar scans. First very general classes of Markov image models were introduced. Then 2-D Kalman filtering methods were presented. Next efficient strip and reduced update filters were presented. The reduced update filter was extended to image restoration. The case of random boundary conditions was treated. Then the 2-D recursive estimation methods were extended to the adaptive multiple-model description and the newly defined doubly stochastic Gaussian random fields. Finally the chapter concluded with work on system identification using spectral and linear predictive models. Ongoing work on the MMRE and doubly stochastic Gaussian parameter identification concluded this section.

Future work should consider the effect of the weak optimality of the reduced update algorithms and develop bounds on their excess mean square error. The additional error due to the approximate reduced update procedure should also be analyzed or bounded. The reduced update procedure could be extended to 2-D ARMA models to obtain increased modeling efficiency. The adaptive multiple-model identification problem should be persued for both noise-free prototype images and for noisy images. The continuously adaptive model parameter estimation should be persued with more powerful identification methods which take the noisy observations into account.

The newly introduced doubly stochastic Gaussian field has many potential applications in image processing. One application is to image source coding as a generalization of the DPCM method.

Acknowledgements. This work was sponsored by the Air Force Office of Scientific Research, Air Force Systems Command. USAF, under Grant No. 77-3361. The United States Government is authorized to reproduce and distribute reprints for Governmental purposes notwithstanding any copyright notation.

The author would like to acknowledge the help of his colleague Professor Howard Kaufman on the adaptive and system identification aspects of this problem. Also our students, Vinay K. Ingle and Assad Radpour, provided significant input to this chapter.

References

5.1 A. Habibi: Proc. IEEE **60**, 878–883 (1972)
5.2 N. E. Nahi: Proc. IEEE **60**, 872–877 (1972)
5.3 S. R. Powell, L. M. Silverman: IEEE Trans. AC-**19**, 8–13 (1974)
5.4 J. W. Woods, C. H. Radewan: IEEE Trans. IT-**23**, 473–482 (1977); IT-**25**, 628–629 (1979)
5.5 M. G. Strintzis: IEEE Trans. AC-**23**, 801–809 (1978)
5.6 A. K. Jain, J. R. Jain: IEEE Trans. AC-**23**, 817–834 (1978)
5.7 M. S. Murphy, L. M. Silverman: "Image Model Representation and Line-by-Line Recursive Restoration", *1976 Conf. Decision and Control*, Clearwater, Florida, Dec. (1976); IEEE Trans. AC-**23**, 809–816 (1978)
5.8 J. W. Woods: IEEE Trans. IT-**18**, 232–240 (1972)
5.9 M. P. Ekstrom, J. W. Woods: IEEE Trans. ASSP-**24**, 115–128 (1976)
5.10 M. P. Ekstrom, R. E. Twogood, J. W. Woods: IEEE Trans. ASSP-**28**, 16–26 (1980)
5.11 A. P. Sage, J. L. Melsa: *Estimation Theory with Applications to Communications and Control* (McGraw-Hill, New York 1971) pp. 89–90
5.12 N. E. Nahi, C. A. Franco: IEEE Trans. C-**21**, 305–311 (1973)
5.13 J. W. Woods, V. K. Ingle: IEEE Trans. ASSP, to appear
5.14 R. A. Wiggins: "On Factoring the Correlations of Discrete Multivariable Stochastic Processes"; Ph. D. Thesis, Department of Geol. and Geophys., M.I.T., Cambridge, MA. (1965)
5.15 W. E. Larimore: Proc. IEEE **65**, 961–970 (1977)
5.16 D. B. Harris, R. M. Mersereau: IEEE Trans. ASSP-**25**, 492–500 (1977)
5.17 A. Radpour, V. Ingle, H. Kaufman, J. Woods: "Adaptive Kalman Type Estimation Applied to Image Processing", 1978 Midwest Symp. Circuits and Syst., Ames. Iowa, August (1978)
5.18 R. S. Lipster, A. N. Shiryayev: *Statistics of Random Processes I. General Theory*, Applications of Mathematics, Vol. 5 (Springer, Berlin, Heidelberg, New York 1977) p. 17
5.19 V. K. Ingle, J. W. Woods: "Multiple Model Recursive Estimation of Images", Proc. IEEE, IASSP'79, Washington, D. C., April (1979) pp. 642–645
5.20 A. S. Willsky: Automatica **12**, 601–611 (1976)
5.21 D. G. Lainiotis: Proc. IEEE **64**, (No. 8) 1126–1198 (1976)
5.22 D. S. Lebedev, L. I. Mirkin: "Smoothing of Two-Dimensional Images Using the 'Composite' Model of a Fragment", Iconics Inst. *Problems in Inform. Trans.* Academy of Science, USSR (1975) pp. 57–62. Reprinted in *Two-Dimensional Digital Signal Processing*, Ed. S. K. Mitra, M. P. Ekstrom. (Dowden, Hutchinson of Ross, Stroudsburg, Pennsylvania, 1978) p. 197–202
5.23 P. A. Ramamoorthy, L. T. Bruton: "Design of Stable Two-Dimensional Digital Recursive Filters and Applications in Image Processing", Proc. 20th Midwest Sympos. Circuits and Systems, Aug. (1977). See also Chap. 3, this volume
5.24 H. Kaufman, A. Radpour: "Adaptive Estimation of Nonstationary Image Processes", Proc. of 1979 John's Hopkins Conf. on Inform. Science and Computers, March (1979)
5.25 J. W. Woods: IEEE Trans. COM-**23**, 845–846 (1975)
5.26 J. B. Anderson, J. B. Bodie: IEEE Trans. IT-**21**, 379–387 (1975)

Additional References with Titles

J. F. Abramatic, F. Germain, E. Rosencher: Design of 2-D separable denominator recursive filters. IEEE Trans. ASSP-**27**, 445–453 (1979)

P. Agathoklis, E. I. Jury, M. Mansour: Evaluation of quantization error in 2-D digital filters. IEEE Trans. ASSP-**28**, 273–279 (1980)

M. O. Ahmad, C. H. Reddy, V. Ramachandran, M. N. S. Swamy: A class of multivariable positive real functions realizable by resistively terminated lossless ladder networks. IEEE Trans. CAS-**26**, 659–662 (1979)

W. E. Alexander, S. A. Pruess: Stability analysis of 2-D digital recursive filters. IEEE Trans. CAS-**27**, 11–15 (1980)

S. Alliney, F. Sgallari: "An Efficient Method for 2-D FIR Filters Approximation", in Proc. EUSIPCO-80, Lausanne, Switzerland, Sept. 16–18, 1980, pp. 397–402

S. A. H. Aly, M. M. Fahmy: "Spatial-Domain Design of 2-D Recursive Digital Filters", in Proc. IEEE Int. Conf. ASSP, Denver, Colorado, April 9–11, 1980, pp. 758–761

N. K. Bose: "2-D Rational Approximates via 1-D Padé Technique", in Proc. EUSIPCO-80, Lausanne, Switzerland, Sept. 16–18, 1980, pp. 409–412

S. C. Bose, R. E. Mortensen: "Two-Dimensional Signal Processing For Vector Random Fields", in Proc. EUSIPCO-80 Lausanne, Switzerland, Sept. 16–18, 1980, pp. 83–88

L. F. Chaparro, E. I. Jury: "Rational Modelling of 2-D LSI Digital Systems", in Proc. EUSIPCO-80, Lausanne, Switzerland, Sept. 16–18, 1980, pp. 417–424

F. Cohen, D. B. Cooper, P. Symosek, H. Elliott: "2-D Image Boundary Estimation by Use of Likelihood Maximization and Kalman Filtering", in Proc. IEEE Int. Conf. ASSP, Denver, Colorado, April 9–11, 1980, pp. 410–413

F. R. Dubrovsky, L. M. Roytman, M. N. S. Swamy: "Some Stability Theorems for Multidimensional Digital Filters", in Proc. 23rd Midwest Sym. Circuits and Systems, Toledo, Ohio, August 4–5, 1980, pp. 467–471

D. E. Dudgeon: "An Iterative Implementation of 2-D Digital Filters", in Proc. IEEE Int. Conf. ASSP, Denver, Colorado, April 9–11, 1980, pp. 741–744

R. Eising: State-space realization and inversion of 2-D systems. IEEE Trans. CAS-**27**, 612–619 (1980)

R. Eising: "Internally Stable Realization of 2-D Transfer Matrices", in Proc. EUSIPCO-80, Lausanne, Switzerland, Sept. 16–18, 1980, pp. 393–396

M. P. Ekstrom, R. E. Twogood, J. W. Woods: Two-dimensional recursive filter design – a spectral factorization approach. IEEE Trans. ASSP-**28**, 16–20 (1980)

N. G. El-Agizi, M. M. Fahmy: Two-dimensional digital filters with no overflow oscillations. IEEE Trans. ASSP-**27**, 465–469 (1979)

A. Fettweis: "Wave Digital Filters – A Review", in Proc. EUSIPCO-80, Lausanne, Switzerland, Sept. 16–18, 1980, pp. 1–6

G. Garibotto: "A New Stability Test for 2-D Digital Filters", in Proc. EUSIPCO-80, Lausanne, Switzerland, Sept. 16–18, 1980, pp. 413–416

T. Hinamoto: A canonical form for a state-space model in 2-D systems. IEEE Trans. CAS-**27**, 710–712 (1980)

E. I. Jury: "Counterexamples in Multidimensional System Theory", in Proc. IEEE Int. Symp. CAS, Houston, Texas, April 28–30, 1980, pp. 196–200

J. F. Kaiser, R. W. Schafer: On the use of the I_0-Sinh window for spectrum analysis. IEEE Trans. ASSP-**28**, 105–107 (1980)

E. W. Kamen: A note on the representation and realization of lumped-distributed networks, delay-differential systems, and 2-D systems. IEEE Trans. CAS-**27**, 430–432 (1980)

J. J. Knab: An alternate Kaiser window. IEEE Trans. ASSP-**27**, 560–561 (1979)

H. Korezlioglu: "On 2-D Innovation Processes", in Proc. EUSIPCO-80, Lausanne, Switzerland, Sept. 16–18, 1980, pp. 187–192

S. Y. Kung, B. Levy, M. Morf, T. Kailath: New result in 2-D system theory, Part II: 2-D state space models. Proc. IEEE **65**, 945–961 (1977)

R. R. Kurth, M. T. McCallig: "2-D FIR Filter Design via Semi-Polynomial Approximation", in Proc. IEEE Int. Conf. ASSP, Denver, Colorado, April 9–11, 1980, pp. 733–736

G. Linnenberg: "Wave Digital Fan Filters: Numerical and Experimental Results", in Proc. EUSIPCO-80, Lausanne, Switzerland, Sept. 16–18, 1980, pp. 15–20

J. H. Lodge, M. M. Fahmy: An efficient l_p optimization technique for the design of 2-D linear-phase FIR digital filters. IEEE Trans. ASSP-**28**, 308–313 (1980)

J. H. Lodge, M. M. Fahmy: An optimization technique for the design of half-plane 2-D recursive digital filters. IEEE Trans. CAS-**27**, 721–724 (1980)

R. M. Mersereau, T. H. Joo, T. C. Speake: "A Comparison of Hexagonally- and Rectangularly-Sampled 2-D FIR Digital Filters", in Proc. IEEE Int. Conf. ASSP, Denver, Colorado, April 9–11, 1980, pp. 729–732

M. Morf, B. C. Levy, S. Y. Kung: New results in 2-D systems theory, Part I: 2-D polynomial matrices. Proc. IEEE **65**, 861–872 (1977)

K. V. V. Murthy, V. Ramachandran, C. S. Gargouv: "On the Generation of Two-Variable Reactance Functions and Matrices Based on the Matrix-Partial Fraction Characterization", in Proc. 23rd Midwest Symp. Circuits and Systems, Toledo, Ohio, August 4–5, 1980, pp. 145–148

T. Nakamura: Construction of stable multidimensional recursive filters via pseudopolynomials. IEEE Trans. ASSP-**28**, 337–339 (1980)

P. N. Paraskevopoulos: Pade-type order reduction of 2-D systems. IEEE Trans. CAS-**27**, 413–416 (1980)

S. A. Rajala, R. J. P. de Figueiredo: "Adaptive Nonlinear Image Restoration by a Modified Kalman Filtering Approach", in Proc. IEEE Int. Conf. ASSP, Denver, Colorado, April 9–11, 1980, pp. 414–417

P. K. Rajan, M. N. S. Swamy: "2-D FIR Filter Design Using Variable Transformations", in Proc. IEEE Int. Symp. CAS, Houston, Texas, April 28–30, 1980, pp. 213–216

B. Raman, V. Ramachandran, M. O. Ahmad, M. N. S. Swamy: "Realization of a Class of Multivariable Network Functions", in Proc. IEEE Int. Symp. CAS, Houston, Texas, April 28–30, 1980, pp. 221–224

C. H. Reddy, P. K. Rajan: "Studies on N-Dimensional Filter Transfer Functions Without Second-Kind Singularities", in Proc. IEEE Int. Conf. ASSP, Denver, Colorado, April 9–11, 1980, pp. 753–757

G. A. Shaw, R. M. Mersereau: "Comparing 2-D Recursive and Nonrecursive Least-Square-Error Approximation Filters", in Proc. IEEE Int. Conf. ASSP, Denver, Colorado, April 9–11, 1980, pp. 762–765

G. L. Sicuranzu: Linear phase approximation for 2-D recursive digital filters. IEEE Trans. ASSP-**27**, 288–291 (1979)

P. Stavroulakis, P. N. Paraskevopoulous: "Reduced Order Feedback Law Implementation for 2-D Digital Systems", in Proc. EUSIPCO-80, Lausanne, Switzerland, Sept. 16–18, 1980, pp. 403–408

B. R. Suresh, B. A. Shenoi: "Pseudorecursion: The 2-D Filtering of Images", in Proc. IEEE Int. Symp. CAS, Houston, Texas, April 28–30, 1980, pp. 225–228

E. Walach, E. Zeheb: Sign test of multivariable real polynomials. IEEE Trans. CAS-**27**, 619–625 (1980)

J. W. Woods, V. K. Ingle, G. Juskovic, R. Hingorani: "Experimental Comparison of Reduced Update Kalman Filters and Wiener Filters of 2-D LMMSE Estimation", in Proc. IEEE Int. Conf. ASSP, Denver, Colorado, April 9–11, 1980, pp. 406–409

Subject Index

Picture Processing and Digital Filtering

Editor: T. S. Huang

2nd corrected and updated edition. 1979.
113 figures, 7 tables. XIII, 297 pages
(Topics in Applied Physics, Volume 6)
ISBN 3-540-09339-7

Contents:
T. S. Huang: Introduction. – *H. C. Andrews:* Two-Dimensional Transforms. – *J. G. Fiasconaro:* Two-Dimensional Nonrecursive Filters. – *R. R. Read, J. L. Shanks, S. Treitel:* Two-Dimensional Recursive Filtering. – *B. R. Frieden:* Image Enhancement and Restoration. – *F. C. Billingsley:* Noise Considerations in Digital Image Processing Hardware. – *T. S. Huang:* Recent Advances in Picture Processing and Digital Filtering. – Subject Index.

Image Reconstruction from Projections

Implementation and Applications

Editor: G. T. Herman

1979. 120 figures, 10 tables. XII, 284 pages
(Topics in Applied Physics, Volume 32)
ISBN 3-540-09417-2

Contents:
G. T. Herman, R. M. Lewitt: Overview of Image Reconstruction from Projections. – *S. W. Rowland:* Computer Implementation of Image Reconstruction Formulas. – *R. N. Bracewell:* Image Reconstruction in Radio Astronomy. – *M. D. Altschuler:* Reconstruction of the Global-Scale Three-Dimensional Solar Corona. – *T. F. Budinger, G. T. Gullberg, R. H. Huesman:* Emission Computed Tomography. – *E. H. Wood, J. H. Kinsey, R. A. Robb, B. K. Gilbert, L. D. Harris, E. L. Ritman:* Applications of High Temporal Resolution Computerized Tomography to Physiology and Medicine.

Springer-Verlag
Berlin
Heidelberg
New York

Digital Picture Analysis

Editor: A. Rosenfeld

1976. 114 figures, 47 tables. XIII, 351 pages
(Topics in Applied Physics, Volume 11)
ISBN 3-540-07579-8

Contents:
A. Rosenfeld: Introduction. – *R. M. Haralick:* Automatic Remote Sensor Image Processing. – *C. A. A. Harlow, S. J. Dwyer, III., G. Lodwick:* On Radiographic Image Analysis. – *R. L. McIlwain, Jr.:* Image Processing in High Energy Physics. – *K. Preston, Jr.:* Digital Picture Analysis in Cytology. – *J. R. Ullmann:* Picture Analysis in Character Recognition.

Nonlinear Methods of Spectral Analysis

Editor: S. Haykin

1979. 45 figures, 2 tables. XI, 247 pages
(Topics in Applied Physics, Volume 34)
ISBN 3-540-09351-6

Contents:
S. Haykin: Introduction. – *S. Haykin, S. Kessler:* Prediction-Error Filtering and Maximum-Entropy Special Estimation. – *T. J. Ulrych, M. Ooe:* Autoregressive and Mixed Autoregressive-Moving Average Models and Spectra. – *E. A. Robinson:* Interative Least-Squares Procedure for ARMA Spectral Estimation. – *J. Capon:* Maximum-Likelihood Spectral Estimation. – *R. N. McDonough:* Application of the Maximum-Likelihood Method and the Maximum-Entropy Method to Array Processing. – Subject Index.

The Computer in Optical Research

Methods and Applications

Editor: B. R. Frieden

1980. 92 figures, 13 tables. XV, 371 pages
(Topics in Applied Physics, Volume 41)
ISBN 3-540-10119-5

Contents:
B. R. Frieden: Introdcution. – *R. Barakat:* The Calculation of Integrals Encountered in Optical Diffraction Theory. – *B. R. Frieden:* Computational Methods of Probability and Statistics. – *A. K. Rigler, R. J. Pegis:* Optimization Methods in Optics. – *L. Mertz:* Computers and Optical Astronomy. – *W. J. Dallas:* Computer-Generated Holograms.

GPSR Compliance

The European Union's (EU) General Product Safety Regulation (GPSR) is a set of rules that requires consumer products to be safe and our obligations to ensure this.

If you have any concerns about our products, you can contact us on ProductSafety@springernature.com

In case Publisher is established outside the EU, the EU authorized representative is:

Springer Nature Customer Service Center GmbH
Europaplatz 3
69115 Heidelberg, Germany

Batch number: 10620163

Printed by Printforce, the Netherlands